Sensors, Cloud, and Fog: The Enabling Technologies for the Internet of Things

Sensors, Cloud, and Fog: The Enabling Technologies for the Internet of Things

Sudip Misra
Subhadeep Sarkar
Subarna Chatterjee

CRC Press
Taylor & Francis Group
Boca Raton London New York

CRC Press is an imprint of the
Taylor & Francis Group, an **informa** business

CRC Press
Taylor & Francis Group
6000 Broken Sound Parkway NW, Suite 300
Boca Raton, FL 33487-2742

CRC Press is an imprint of Taylor & Francis Group, an Informa business

Printed on acid-free paper

International Standard Book Number-13: 978-0-3671-9612-7 (Hardback)

Library of Congress Cataloging-in-Publication Data

Names: Misra, Sudip, author. | Sarkar, Subhadeep, author. | Chatterjee, Subarna, author.
Title: Sensors, cloud, and fog : the enabling technologies for the internet of things / Sudip Misra, Subhadeep Sarkar, Subarna Chatterjee.
Description: Boca Raton : Taylor & Francis, a CRC title, part of the Taylor & Francis imprint, a member of the Taylor & Francis Group, the academic division of T&F Informa, plc, 2019. | Includes bibliographical references.
Identifiers: LCCN 2019020922 (print) | ISBN 9780367196127 (pb : acid-free paper)
Subjects: LCSH: Internet of things.
Classification: LCC TK5105.8857 .M57 2019 (print) | LCC TK5105.8857 (ebook) | DDC 004.67/8–dc23
LC record available at https://lccn.loc.gov/2019020922
LC ebook record available at https://lccn.loc.gov/2019981504

Visit the Taylor & Francis Web site at
www.taylorandfrancis.com

and the CRC Press Web site at
www.crcpress.com

Dedicated to Our Families

Contents

Foreword

This book brings forth new dimensions to technological challenges and advances towards the transformation of the Internet of Things (IoT), from a mere concept to established and deployable technology. The authors focus on two key enabling technologies of IoT: the sensor-cloud and fog computing.

This book provides an in-depth understanding of the IoT technology for students, researchers, professionals, and practitioners in the field. The authors highlight the research and the technological challenges, which one faces while translating this concept of IoT into a practical, technologically feasible, and business-viable solution. The book discusses two principal enabling technologies for IoT – the sensor-cloud and fog computing – which serve as backbones of sensing and computing in IoT. Each chapter of the book has multiple illustrations and examples supported by relevant figures and diagrams. Experimental results from existing literature are presented in the book. Every chapter includes a summary of the chapter's discussions and observations and an exercise in the form of a set of conceptual questions and programming examples. Hints are provided for advanced problems. There is a very balanced mixture of theoretical insights and application-oriented analysis, making this book unique.

Professor Sudip Misra is widely known for the books he has co-authored and edited with reputed publishing houses such as Cambridge University Press, Wiley, and Springer. The work of Dr. Chatterjee on sensor-cloud has been extensively referenced. Dr. Sarkar's publications in fog computing are highly cited by the research community.

I hope that this text will serve as a wealth of timely knowledge for the students and researchers in the domain.

Professor Bharat K. Bhargava, *Fellow of IEEE*
Department of Computer Sciences, Purdue University

Preface

Overview

With the inception of the emerging technology of the Internet of Things (IoT), it is anticipated that by the end of 2020, billions of sensor-enabled devices will be connected to the Internet. The global market of IoT is expected to observe a revenue growth of around 300 billion. Therefore, it is intuited that a very large number of sensors and applications will be involved in the functioning of IoT, leading to enormous growth in the number of IoT users. Since its conceptualization, IoT has been envisioned as a technology that is highly reliant on sensor-based data acquisition and processing of the data in real time.

Traditional wireless sensor networks (WSNs) were envisioned to play a significant role for sensor-based data acquisition in IoT. However, the principal problems that persist relating to WSN-based data acquisition are (a) WSNs are expensive and are usually single user-centric, (b) they can be used only after they are purchased and deployed by the end-users, who are also responsible for their maintenance, and (c) there is difficulty in scaling and customization of WSNs. Therefore, the reachability of WSNs for the common mass is restricted. In order to address these problems, sensor-cloud infrastructure was proposed and envisioned as a substitute of WSNs. However, in classical cloud computing, processing and storage of data happen only within the cloud data centers. Given the huge volume of data generated by the IoT devices, the classical cloud-computing framework experiences a major network bottleneck, resulting in high service latency and poor quality of service (QoS). To address this problem, another very recent computing paradigm emerged in the form of fog computing. Fog computing is a distributed computing paradigm that empowers network devices at different hierarchical levels with various degrees of computational and storage capability. These devices are equipped with "intelligence", which allows them to examine whether an application request requires the intervention of the cloud-computing tier. The idea is to serve requests that demand real-time, low-latency services by the fog computing devices, the connected work stations, and the small-scale storage units.

Together, these two technologies act as major enablers towards the realization of IoT. Therefore, while this book is intended for masters- and doctoral-level students interested in advanced courses on Internet technologies, it is also expected to be highly popular in the researcher community. This book attempts to provide a thorough overview of IoT, the key enablers of the technology, the inter-relationship of the components, and the various software and hardware models used to realize IoT.

Organization of the Book

This book is organized to help the readers walk through the basic concepts of IoT, build a foundation on the technology, and eventually understand the advanced concepts. The book comprises three parts – in the first part, we present the evolution of computing technologies, with a focus on the distributed computing infrastructures, and introduce IoT. The second part of the book is focused on sensor-cloud technology, as we begin with an introduction to this new paradigm and then present its data management, networking, and pricing solutions. We also emphasize how this new technology facilitates realization of the IoT. In the final part of the book, we present, as a computational framework, the fog-computing paradigm. Following the introduction of this new breed of computing framework, we present its different IoT-related application domains and present the architecture of fog computing. We conclude this part by providing a detailed analysis of how and why fog computing holds the key as the computing infrastructure for the IoT. This particular way of organizing the chapters in the book helps the readers to develop an understanding of the IoT (and gives them a brief idea about WSNs and cloud computing). Then, based on this foundation, we provide the different challenges in realizing IoT in practice and present the sensor-cloud and fog computing paradigms. We clearly put forward the salient features of both technologies and show how these novel technologies are set to transform the world of IoT.

Organization of the Chapters

In Chapter 1, we discuss the history and evolution of cloud computing. We start by describing the different cloud deployment and service models, followed by a presentation of computation that takes place within cloud platforms. We specifically address the topic of resource management, virtualization, and green computation within cloud servers. Finally, we present the cloud applications that are widely used these days.

In Chapter 2, we present the fundamentals of sensor networks: their background and evolution, the architectural design of sensor nodes, sensor-based applications, and the existing open research challenges.

Chapter 3 introduces IoT by defining the key components highlighting its defining characteristics. We also describe the details of the operations within the IoT middleware and familiarize readers with various IoT application domains, such as healthcare, smart city, telecommunication, and supply chain management.

Chapter 4 presents the concept of sensor-cloud including its background, sensor virtualization, and applications. The chapter also discusses several experimental results that support the shift of the sensing paradigm from WSNs to sensor-cloud.

In Chapter 5, we discuss the details of data flow within sensor-cloud. We illustrate how virtual sensors are formed and how data are managed through internal caching mechanisms within a sensor-cloud platform.

Chapter 6 brings forth the pricing mechanisms within sensor-cloud. It discusses two different pricing models and their respective implementations. The chapter also highlights the concept of data center networking within sensor-cloud. We specifically present the problem of selecting data centers optimally to maintain application Quality of Service (QoS).

Chapter 7 presents how the emergence and evolution of sensor-cloud directly facilitate the growth of IoT through its architecture, functionalities, and life cycle.

Chapter 8 introduces the concept of fog computing. It elaborates upon the limitations of cloud computing and narrates the power of fog computing to overcome them.

Chapter 9 presents fog-based applications in the domains of healthcare, smart cities, and other emerging application scenarios.

In Chapter 10, we discuss the fog computing architecture and describe its key components. We also present the mathematical model of fog architecture and provide insights on application-specific architectures.

In Chapter 11, we discuss how fog computing can serve as a green computing paradigm in the context of IoT. We strengthen this discussion with the help of a case study to demonstrate the network topology and the various parameters involved.

Finally, Chapter 12 discusses the security aspects of IoT and present open research issues.

Target Audience

This book will be helpful to students, researchers, and practitioners working in the field, who seek to acquire insights in the topics of IoT, sensor-cloud, and fog computing. While a number of books relate to IoT, the existing books primarily focus on the fundamentals of the IoT and discuss introductory concepts, applications, advantages, and principles of IoT. A few advanced books in the domain aim at helping software engineers, web designers, and product designers who work on designing products for IoT. However, no books highlight the key enabling technologies for IoT, which are particularly of interest for students and researchers specializing in the topic. Presently, only a handful of edited books and a few book chapters address the topics of sensor-cloud and fog computing. However, going by the citations received by our published articles, there is a great interest in the research community in the aforementioned topics. From a lecturer's point of view, this book could be the basis of several advanced courses and a fundamental guide for the students. Lastly, we envisage the book as highly relevant for professionals and practitioners in the field who want to enhance their knowledge by developing an understanding of recent technological breakthroughs.

Acknowledgments

We would like to thank our friends and families for their endurance and support. We would also like to thank our colleagues who have always been a source of inspiration, guidance, and support. We would especially like to thank Aishwariya Chakraborty and Chandana Roy who gave their valuable time to proofread. Finally, we thank the publishers for accepting our book proposal and making it see the light of the day.

About the Authors

Sudip Misra is a Professor and Abdul Kalam Technology Innovation National Fellow in the Department of Computer Science and Engineering at the Indian Institute of Technology Kharagpur. He received his Ph.D. degree in Computer Science from Carleton University, in Ottawa, Canada. His current research interests include Wireless Sensor Networks and Internet of Things. Professor Misra is the author of over 300 scholarly research papers. Dr. Misra has published over a dozen books including the popular ones such as the *Principles of Wireless Sensor Networks* (Cambridge University Press), *Smart Grid Technology* (Cambridge University Press), *Network Routing* (Wiley), and *Opportunistic Mobile Networks* (Springer). He has won *10 research paper awards* in different conferences and journals. He was awarded the *IEEE ComSoc Asia Pacific Outstanding Young Researcher Award* at IEEE GLOBECOM 2012, California, USA. He was also the recipient of several academic awards and fellowships such as the *Faculty Excellence Award* (IIT Kharagpur), *Young Scientist Award* (National Academy of Sciences, India), *Young Systems Scientist Award* (Systems Society of India), *Young Engineers Award* (Institution of Engineers, India), *(Canadian) Governor General's Academic Gold Medal* at Carleton University, the *University Outstanding Graduate Student Award* in the Doctoral level at Carleton University and the *National Academy of Sciences, India – Swarna Jayanti Puraskar* (Golden Jubilee Award). He was awarded the Canadian Government's prestigious *NSERC Post Doctoral Fellowship and the Humboldt Research Fellowship* in Germany. Dr. Misra has been serving as the *Associate Editor* of the IEEE Transactions on Mobile Computing, IEEE Transactions on Vehicular Technology, and IEEE Systems Journal, Pervasive and Mobile Computing (Elsevier), International Journal of Communication Systems (Wiley), and IET Communications Journal. He is the *Fellow* of the National Academy of Sciences (NASI), India, the Institution of Engineering and Technology (IET), UK, the Institution of Electronics and Telecommunications Engineering (IETE), India, the Royal Society of Public Health (UK), and the British Computer Society (UK).

Subhadeep Sarkar earned his PhD from the Indian Institute of Technology, Kharagpur. He is currently at Boston University. His current research interests include data storage systems, access methods, and data management systems for large-scale data management systems including cloud/fog computing infrastructures. Dr. Subhadeep was selected as one of "Most Qualified Young Scientists" in Computer Science to attend the Heidelberg Laureate Forum (2016). He has served in many program committees for reputed conferences and is also a reviewer to several journals. Dr. Subhadeep is also a member of the ACM SIGMOD Reproducibility Committee (2018). This book project was initiated when he was at the Indian Institute of Technology, Kharagpur.

Subarna Chatterjee completed her doctoral studies at the Indian Institute of Technology (IIT) Kharagpur. She is currently affiliated with Harvard University. She is a Google Anita Borg Scholar from the Asia Pacific region (2015) and is a Facebook GraceHopper Scholar (2016). Her current research interests include big data stream processing, cloud computing, and Internet of Things. This book project was initiated when she was at IIT.

I

INTRODUCTION

Chapter 1

History and Evolution of Cloud Computing

1.1 Introduction

Cloud computing is an information technology paradigm that ensures on-demand delivery of services in terms of software, platform, and infrastructure. According to NIST [1], cloud computing is defined as "a model for enabling ubiquitous, convenient, on-demand network access to a shared pool of configurable computing resources (e.g., networks, servers, storage, applications, and services) that can be rapidly provisioned and released with minimal management effort or service provider interaction."

We have always wondered about determining the time when cloud computing was invented. Generally, we consider that the concept first emerged in the 21st century [2]. However, this statement is partially true, as the concept of cloud computing has existed and been in use for years. In the 1950s, when people worked on mainframe computers, they accessed the machines using dumb terminals [2]. Due to the huge costs and overhead associated with the gigantic machines, it was not possible to provide a machine for each employee in an organization. Thus, the idea of sharing access to a single resource emerged. Later in the 1970s, the concept of virtual machines (VMs) emerged. With the help of software for virtualization, computing environments could be created in isolation within mainframe servers. This enabled the existence of multiple operating systems within the same computing environment. In 1990s, telecommunications companies provided virtualized private networks instead of dedicated point-to-point data connections.

We will first discuss the classification of cloud computing and then present the different characteristic aspects of the paradigm.

1.1.1 Classification of Cloud Computing

Computing clouds can be classified by their location or the services they offer. In this subsection, we illustrate the different cloud-computing models based on their deployment patterns and services offered.

1.1.2 Cloud Computing Deployment Models

1. **Public**: In this type of cloud deployment [3–7], the entire computing infrastructure is available to the customers publicly over the Internet, as shown in Figure 1.1. Some examples of public cloud deployments are Amazon Web Services (AWS), Microsoft Azure, and Google Cloud Platform. For example, end-users can obtain the services of VMs, cloud storage, or application processing; the services can be free or payable based on the use. It is a multi-tenant virtualized environment for sharing and provisioning of resources. It demands a high bandwidth and large storage for transmitting services and redundant replication of stored data.

The primary advantages of public cloud are as follows:

- It mitigates the requirements of investment and periodic maintenance of on-premise IT resources [2,8].
- There is less resource wastage, as customers can simply pay for the consumed resources.

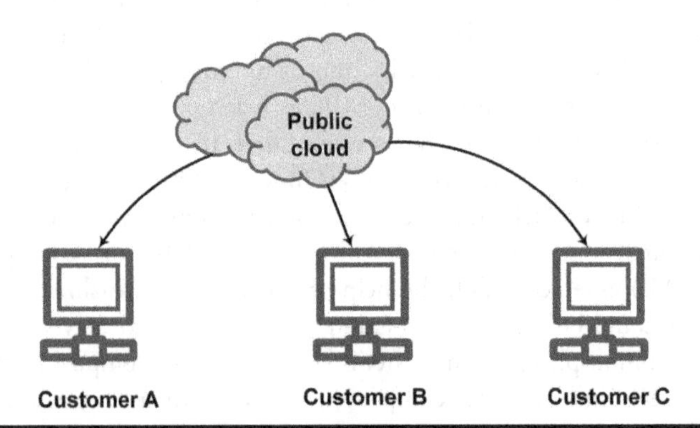

Figure 1.1 Public cloud deployment model

2. **Private**: In private cloud deployment models [9–12], the infrastructure is solely deployed and maintained within the organization, but it has a proprietary architecture [6], as depicted in Figure 1.2. It fulfills the demands and requirements of services and applications internal to the organization and are not for the public. Private clouds comprise the resources and infrastructure of an organization's data centers. Typical examples of private clouds are Amazon Elastic Compute Cloud (EC2), IBM's Blue Cloud, Sun Cloud, Google App Engine, and Windows Azure Services Platform.

The primary advantages of private cloud are as follows:

- It has all the advantages of public cloud infrastructures.
- The company can maintain full-fledged control of the resources and the hardware layer of the cloud servers.

3. **Hybrid**: A hybrid cloud deployment model [13–16] is a combination of public and private cloud deployments, as shown in Figure 1.3. For example, the organization may use the public cloud for data storage and processing and use the private cloud for deploying and executing legacy applications. For example [17], if the demands of a company's cloud services typically increase at a certain time of the year, it can choose to support the heavier traffic by using a hybrid strategy. However, hybrid cloud solutions might be expensive, so they are to be used wisely and strategically.

4. **Community**: Community cloud is defined as "a cloud service model that provides a cloud computing solution to a limited number of individuals or organizations that is governed, managed and secured commonly by all the participating organizations or a third party managed service provider" [18].

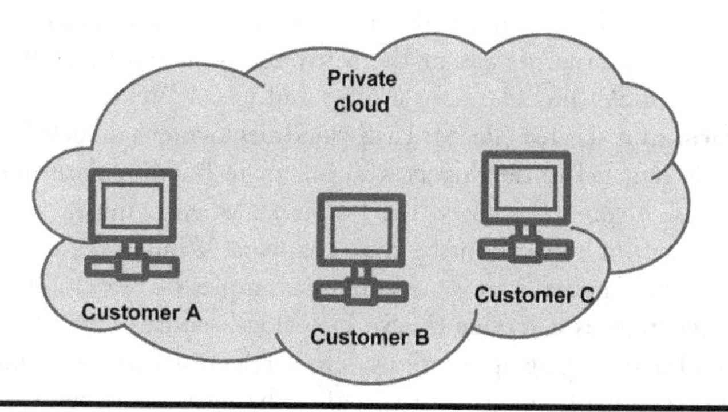

Figure 1.2 Private cloud deployment model

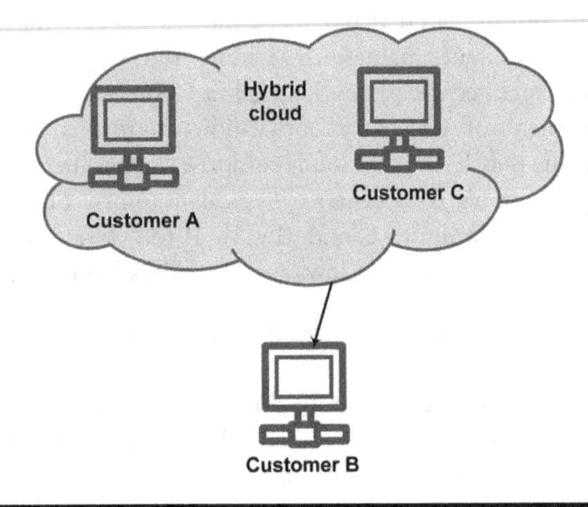

Figure 1.3 Hybrid cloud deployment model

These are the hybrid forms of private clouds [19–22] which are developed for a specific customer, as indicated in Figure 1.4.

1.1.3 Cloud Computing Service Models

Cloud offers a wide range of services from storage to infrastructure, platform, software, and resources. The major categories of cloud services are as follows:

- **Software as a service (SaaS)**: In SaaS cloud models [23–26], providers offer software services in terms of complete products. The end-users are abstracted from the back-end processing and provisioning mechanisms, and they simply do not need to install or maintain the software. SaaS is provisioned in a way to benefit cloud customers to access and utilize any software over the Internet without worrying about the specification, configuration, or internal details of the software, as shown in Figure 1.5. Some typical examples include Google Apps and Cisco WebEx.
- **Platform as a service (PaaS)**: PaaS cloud deployment models [14,27,28] are generally targeted to developers who intend to build applications on specific platforms. Figure 1.5 shows that PaaS enables easy, quick, and convenient management of applications by the developers. Windows Azure, Google App Engine, and Apache Stratos are popular examples of PaaS models.
- **Infrastructure as a service (IaaS)**: IaaS cloud services [30–32] are primarily utilized for managing applications. The services of IaaS cloud include access and utilization of storage, virtualization, hardware, and network resources, as shown in Figure 1.5. Some common examples of IaaS are AWS, Cisco Metapod, and Microsoft Azure.

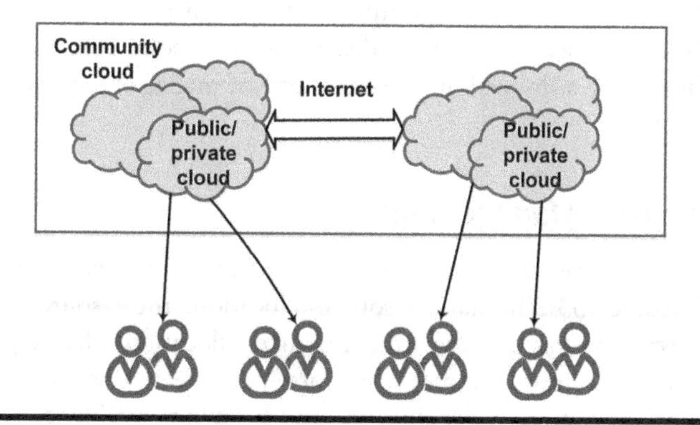

Figure 1.4 Community cloud deployment models

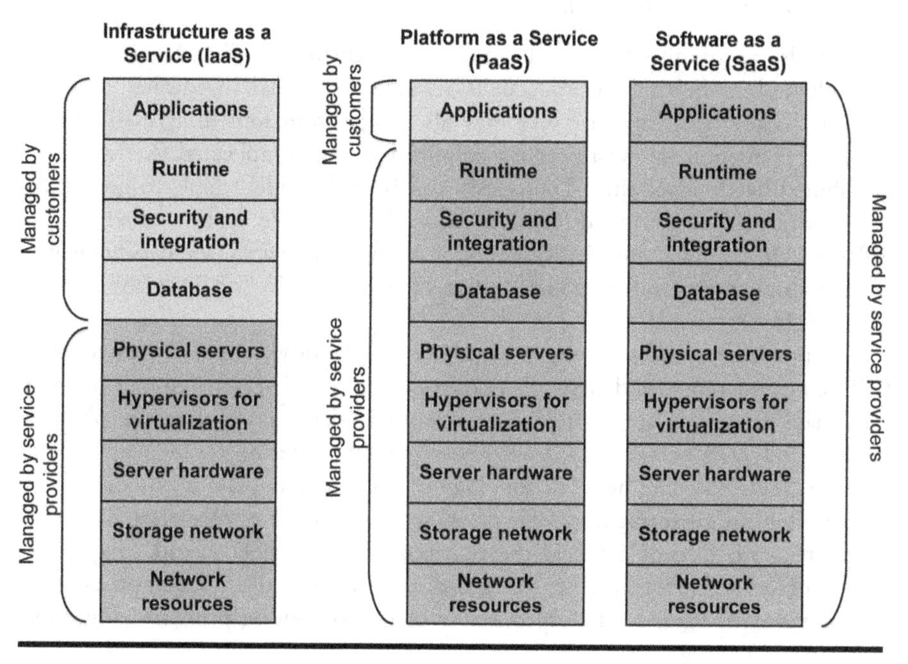

Figure 1.5 Cloud service models [29]

1.2 Computation in Cloud

Computation in cloud is a significant and interesting problem even today, as it addresses several technological challenges related to resource management, virtualization, reducing performance overhead, and ensuring energy efficiency. It also addresses a fascinating variety of multiple optimization problems, most of which are of the NP-hard type. We begin by discussing the resource management issues in cloud, followed by a discussion on the aspects of management and monitoring of virtualization.

1.2.1 Resource Management

In cloud computing, resource management is classified into two different types – static and dynamic [33]. In static resource allocation, the resource requests are known *a priori*, whereas for dynamic resource allocation, the requests arrive dynamically at application runtime. Therefore, resource management in cloud is a very complex process, as it depends on user behavior, workload patterns, and system load. We separately discuss the issues in the following subsections.

1.2.1.1 Workload Models

Workloads are submitted to cloud servers by the customers. Workloads comprise multiple tasks requiring heterogeneous resources in small, moderate, or large amounts. The granularity of a task may also range from something as simple as an HTTP request or word count application to complicated statistical or distributed batch processing. These tasks can be independent (able to be carried out in distinct computing threads) or dependent (unable to be carried out in isolation). In complex models, where the tasks are interrelated, the dependency is studied using two popular models – Bag of Tasks (BoT) and Directed Acyclic Graph (DAG).

The type of resources serving a workload hugely vary – from the underlying CPU cores, memory, and network resources to physical servers, containers, and even data centers. Data centers may comprise identical or different physical servers, and each server may have heterogeneity in terms of the component physical resources. The network topology of data centers plays a crucial role in determining the performance of the cloud servers in serving workloads. The data center network directly impacts the transmission time, delays, and congestion. The problem is more intense for geo-distributed cloud that comprises geo-distributed data centers. The objectives of cloud computing primarily target the following [33]:

Performance: The performance of cloud services is measured in terms of the response time of end-users, turnaround time of applications, and the overall waiting time. Further, service time is the measurement of the duration of time a cloud service is available, an important metric that indicates the "busy"-ness of a service provider.

Finance: As cloud follows a pay-per-use model, customers are charged only for the units of services they consume. However, in the process of charging, customers tend to choose the service provider who charges the lowest price. Therefore, there is a trade-off between the price charged by the provider and the overall profit.

Environment: The impact on the environment is a concerning issue these days due to the massive emission of CO_2 from powerful computing servers. Therefore, minimizing the harmful impact on the environment is critical. The parameters affecting performance in this regard are usage of electrical energy, cooling costs, peak power consumption, and CO_2 emission.

Other: Other auxiliary factors that contribute towards performance objectives are reliability, security, data privacy, and legal compliance.

Now, we discuss the different aspects of resource management (see Figure 1.6).

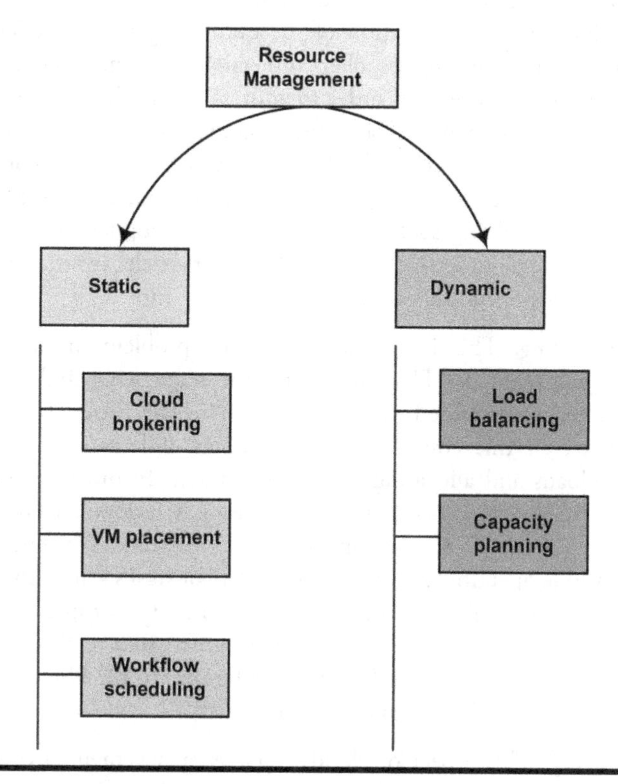

Figure 1.6 Resource management in cloud computing [33]

Cloud Brokering: Often, cloud workloads are submitted by third-party brokers. The primary goal of brokers is to match resources to a job or demand request for multiple customers. The problem concerns minimizing the cost to the customer while allocating the optimal resources. There are several approaches to this classic problem. The first approach is based on the hypothesis that the provider infrastructure comprises homogeneous and infinite resources. Thus, it is possible to allocate VMs to distinct tasks of a job. In order to address the problem, Genetic Algorithm is very often used. For example, Frey *et al.* [34] proposed a simulation-based genetic algorithm CDOXplorer and studied the feasibility, competitiveness, and scalability of the solution. The authors studied twelve VM instance types of Amazon EC2 and five Microsoft Windows Azure units with a running time less than 30 seconds. The experimentation depicts the solution becoming ready to respond before the running time period. There are other approaches to solving the problem of cloud brokering by dealing with either a fixed amount of resources or a fixed number of service providers.

VM Placement: Placing VMs within the physical servers of a cloud data center concerns the allocation and utilization of the underlying resources. This can be thought of as another mapping problem, in which the requirement is to map VMs to physical server clusters. In order to address this problem, it is imperative to understand the requirements of both the customers and the provider in terms of the migration time, resource availability and utilization, power consumption, and service latency. Once these requirements are determined, there are several typical approaches to solving such problems, the most popular based on genetic algorithms (GA), single-objective or self-adaptive particle swarm optimization (PSO), ranking chaos optimization (RCO), and many more.

Workflow Scheduling: This is another interesting problem in the domain of cloud resource management. This constitutes the sequencing and allocation of multiple precedence-constrained tasks represented in the form of a DAG. The principal objective in this context is the reduction of latency and cost, while organizing workloads and allocating resources to them. In order to address such problems, some scientific works focus on proposing solutions encoded with the permutation of sub-tasks. Several operators can be used for encoding. For example, mutation operator modifies the order of tasks, and the crossover operator interchanges the position of tasks. Sometimes, tasks can be organized in terms of execution priorities. The approaches taken to resolve the mathematical optimization include Improved Differential Evolution (IDE), PSO, multi-parent crossover operator (MPCO), and many more.

Capacity Planning: This aspect of cloud resource management involves prediction and anticipation of workload fluctuations and dynamism of resource requirements. Therefore, the primary objective is to satisfy the user requirements in terms of application QoS with minimized cost. In this scenario, while under-provisioning of resources directly affects the application QoS, over-provisioning of resources leads to not only resource wastage, but also the difficulty of serving other applications in a finite-resource situation. For addressing this issue, existing research works stochastically model a cloud system. The solution is selected using uniform probability mass function. Some works [18,34] focus on the self-tuning of optimization parameters (e.g., throughput, latency, and cost), which are monitored dynamically using a QoS monitor. Some works also focus on resource prediction and management using multiplayer online games. Artificial neural network (ANN) is majorly used for prediction of resources, and the model is trained based on the QoS monitor.

Load Balancing: This aspect focuses on optimally dispatching the incoming user requests to the pool of available resources. Similar to capacity planning, the goal is to meet the requirement of application QoS by minimizing cost. This can be visualized as honey-bee behavior, in which bees are associated with the overloaded tasks, and the destination of bees refers to the VMs that are currently under-utilized and hence can accept and serve more requests. Solution approaches to these problems include radial basis functions ANN (RBFs) and adaptive neuro-fuzzy inference system (ANFIS). Some works employ pattern recognition to detect and classify application requests with similar workloads or QoS requirements.

Optimized allocation of cloud resources induces greater research attention, especially with the contemporary data explosion. Scalability of resources at runtime is still limited, as the data sets are trained offline with a specific training method. Therefore, there is always scope for enhancing and optimizing the existing resource allocation approaches [45].

1.2.2 Virtualization

Referring back to Figure 1.5, we observe that IaaS cloud provides the maximum flexibility of services, as developers can design, allocate, and play with their own requirements of VMs. One of the main advantages of using IaaS cloud, other than the wide spectrum of access to cloud resources, is that the customers are charged only for the leased resources and the duration of time of using the resources. However, the increase in the number of tenants leads to contention in adopting these models, as the latter are primarily based on resource sharing. Therefore, performance of the VMs is a critical issue that demands to be addressed and taken care of.

The performance of VM overhead can be studied under three real-life representative scenarios – single server virtualization, single data center, and several geo-distributed data centers [35]. In a single server virtualization scenario, contentions for the underlying CPU cores, memory, and network resources are observed. Several approaches focus on allocation of resources in isolation; some focus on performance modeling and advanced resource isolation techniques.

Modeling and addressing performance overhead of VMs within a data center or across multiple geo-distributed data centers are extremely challenging tasks; they cannot be supported by traditional resource isolation techniques. In these scenarios, the magnitude of the overhead blows up because of the large number of tenants, the heterogeneous demands, the fluctuating workloads, the variation of QoS, and most importantly, scarcity of resources. We will now discuss the different causes of VM overhead and the ways to mitigate them for all the three representative scenarios.

1.2.2.1 Single-Server Virtualization

Hypervisors play a significant role in resource management and allocation of co-located VMs in a single physical server. The state-of-the-art hypervisors, such as Microsoft *Hyper-V* and Citrix *XenServer* target the sharing of CPU cores, disk memory, and main memory, as shown in Figure 1.7. These hypervisors employ their own CPU schedulers and partition memory based on the requirements and

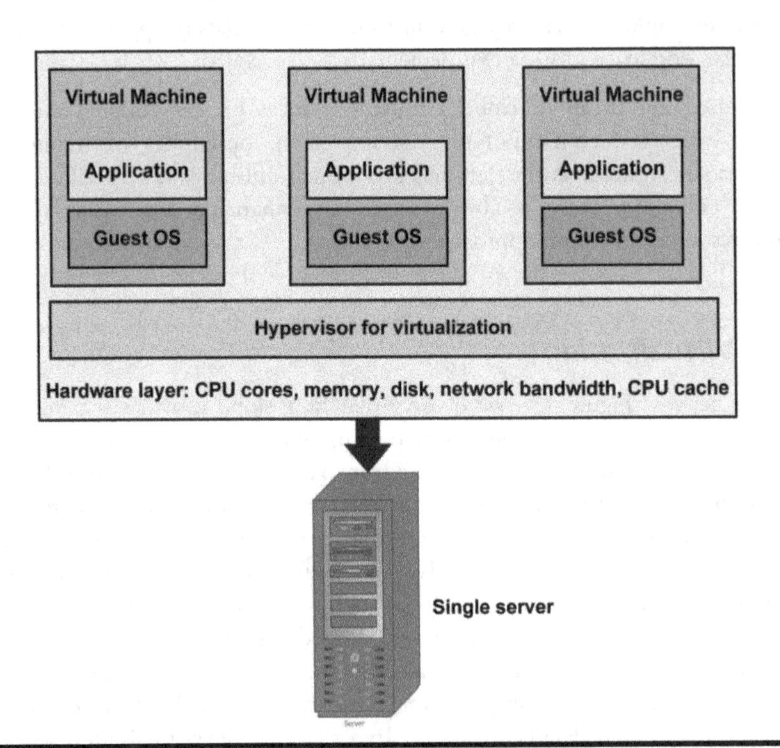

Figure 1.7 Single-server virtualization

capacities. However, resources such as network and disk I/O bandwidth and CPU cache are not well controlled for allocation in isolation [35]. This contention of cache and bandwidth generates overhead for management of VMs in single server virtualization scenarios.

1.2.2.2 Single Data Center

In a single large data center, the overhead of virtualization is manifested in scenarios when a customer uses multiple VMs for a large job. In such a case, multiple VMs residing on several physical machines are allocated to serve the job request, as indicated in Figure 1.8. To continue the normalcy of operation within these VMs, certain routines are executed by the service provider periodically, such as inter-VM data migration, live VM migration, and periodic checkpointing and snapshotting.

Live VM migration is extremely crucial to prevent an application from undergoing a long down time, which, in turn, affects the QoS. From the work of Jung *et al.*, we observe that live migration of VMs (with 256MB RAM supported by the Xen Dom-0 hypervisor) executing RUBiS-1 and RUBiS-2 applications encounters an increase in the response time by 211%, as compared to the performance before migration [36]. Thus, migration of multiple VMs is expected to further increase the overhead and thereby complicate the overall virtualization and resource management within the data center.

From the perspective of fault-tolerance of VMs, the service provider schedules periodic checkpointing of VMs. However, these routines generate significant network traffic, which consumes huge bandwidth. Additionally, service providers deploy storage area networks (SANs) for storing data coming from the data centers. However, multiple concurrent points of access to SANs create contention, which affects the I/O throughput and latencies.

1.2.2.3 Multiple Geo-Distributed Data Centers

When multiple data centers are geo-distributed yet coordinated to work in a collaborative manner, WANs form an indispensable way to connect them for migration of VMs and access to SANs. In such cases, it is possible to migrate VMs from over-loaded data centers to the under-utilized ones for load balancing purposes. This is particularly useful in hybrid cloud-based organizations that decide to transfer VMs from public to private cloud, based on circumstances. Figure 1.9 shows that such VM migrations over WAN are different from the aforementioned VM migrations within a single data center or a single server. This is because, in addition to the transfer of data related to the VM memory, it is imperative to transfer data representing the disk image and network connections. This creates additional performance overhead for geo-distributed data centers.

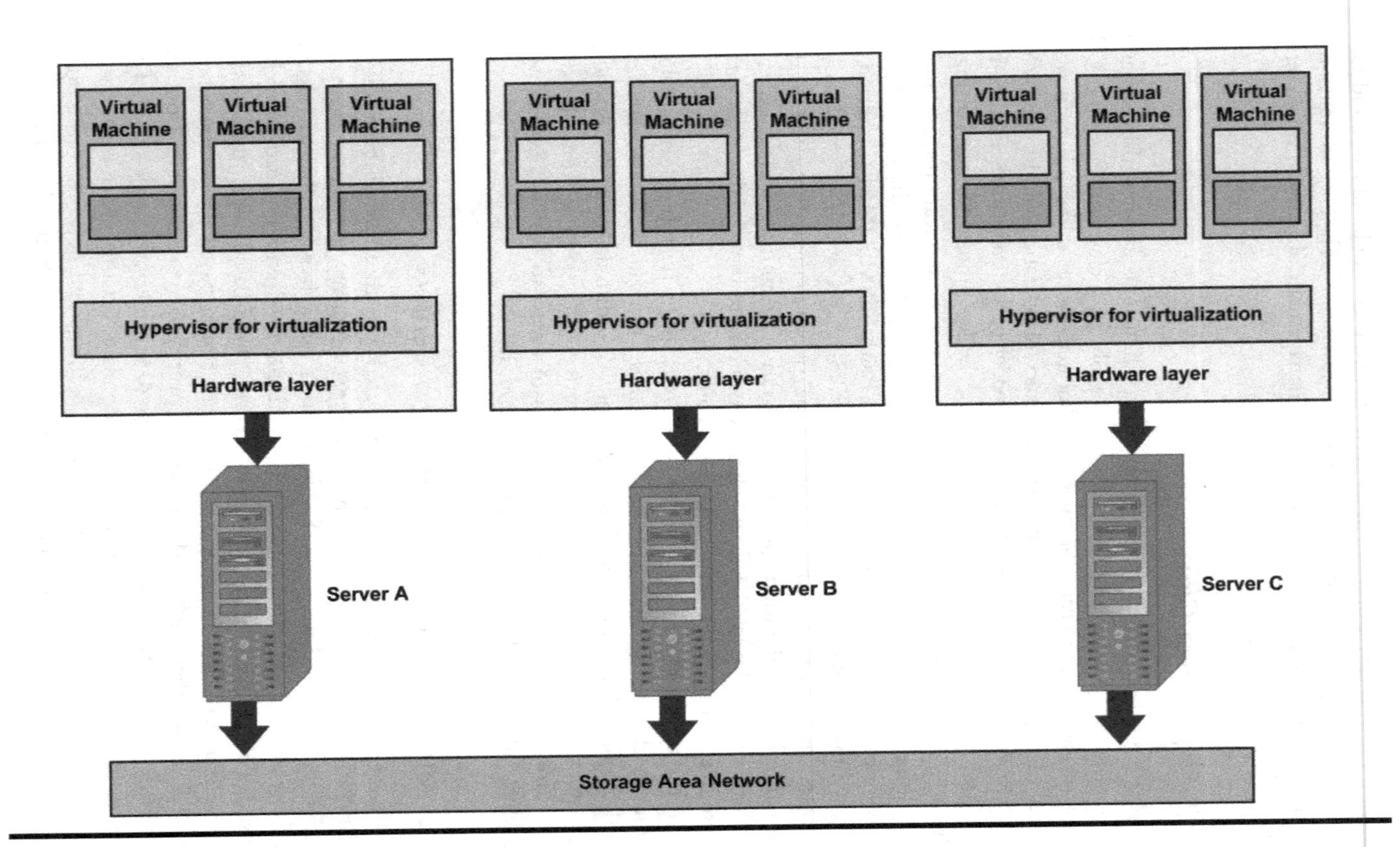

Figure 1.8 Virtualization within a single data center

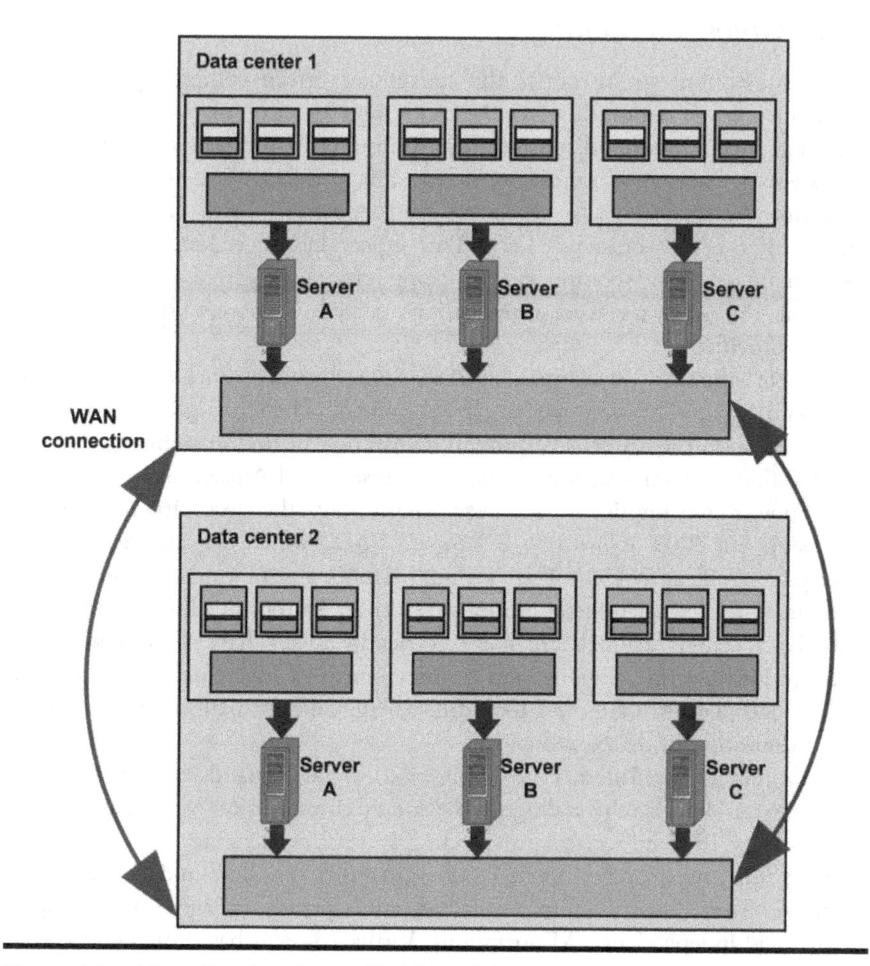

Figure 1.9 Virtualization in geo-distributed data centers

The problem of virtualization under the afore-mentioned categories is dealt with several performance modeling, optimization, and prediction techniques. Correlation and regression analysis are widely used for the estimation of workload fluctuations and VM performance overhead. However, statistical methods require pre-execution of the algorithms to deduce inferences. This is not always practically viable because in real-life scenarios, heterogeneity is very common for workload, applications, and resource demands. Therefore, investigation and modeling of the performance of VMs is an ongoing challenge.

1.2.3 Green Computing

In this subsection, we introduce the concept of "green computing". Murugesan defines green computing as "the study and practice of designing, manufacturing, using and disposing of computers, servers and associated subsystems – such as monitors, printers, storage devices and networking and communications systems – efficiently and effectively with minimal or no impact on the environment" [37]. This aspect has increased significance in the domain of cloud computing, as computations in cloud are complex in terms of the software used, applications served, resources consumed, and services provided.

In 1992, the US Environmental Protection Agency launched Energy Star, a program to support and promote climate control. The term "green computing" was coined soon thereafter. The primary motivation behind introducing the term was that big companies such as Google, Microsoft, and Amazon have huge costs due to power consumption. It has been reported that the electricity consumption of Google in 2009 is worth 38 million USD, which, of course, increases every year significantly [38]. This not only implies a significant reduction in the profit margin of the cloud service providers, but also ensures huge emissions of CO_2. Data centers as such are not eco-friendly. They have a massive carbon footprint for powering, cooling, and computation [39]. Therefore, green computing in cloud platforms attempts to address energy-efficient performance modeling and optimization of data centers.

In the era of the Internet of Things (IoT), the demands of customers are increasing rapidly, thereby scaling up the energy consumption of the data centers. Thus, energy-efficient service provisioning is very challenging yet essential for ensuring the sustainability of IoT. Towards this, Nathuji and Schwan [40] proposed VirtualPower – an approach for online power management to support isolated and independent VM operations. VirtualPower is based on Xen hypervisor, which provisions several system-level abstractions and achieves a 34% reduction in power consumption. Kusic *et al.* [41] proposed a dynamic resource-provisioning framework taking into account the switching costs incurred in provisioning VMs. The work achieves a 26% power consumption reduction, while still maintaining application QoS.

For energy-efficient computation in microprocessors, new materials [39] have been introduced to reduce energy consumption. For example, thick layers of SiO_2 gates are now replaced by layers of hafnium-based oxide, resulting in reduced generation of gate tunneling currents without affecting the performance of the transistor [42,43]. For reduction of power during computation, thermal-aware scheduling algorithms are introduced with genetic algorithm and quadratic programming. Simultaneously, chilled-liquid coolers are also used within data centers with microchannel heat sinks.

1.3 Cloud Applications

Following are various applications in which cloud computing plays an indispensable role and is commonly used in our daily lives.

Storage: The storage of digital data is crucial. Because of the data explosion around us, this requirement is increasing continuously. Popular online cloud storage providers such as Google and Dropbox sprang up to meet the needs of billions of people across the globe. On one hand, Google has revolutionized the way we store emails, organize media data (photos and videos), store personal data, and maintain schedules and appointments. On the other hand, Dropbox enables easy online storage with versioning facilities and convenient data sharing.

Social networking: The Global Digital Report 2018 states that the number of social media users across the world is 3.2 billion with an annual growth rate of 13% [44]. The most popular social networking service providers are Facebook, Twitter, and Pinterest. The social network users utilize the private cloud of the service providers to store a gigantic volume of personal and media information. Such clouds are subject to continuous learning for being adaptive to customized personal data analytics.

Healthcare: Dell's Secure Healthcare Cloud and IBM Cloud enable ubiquitous monitoring of patients by storing and sharing massive volumes of health data for a large number of global patients. The health conditions of patients are updated, tracked, and diagnosed remotely through the provisioned cloud services. Cloud servers are also deployed within the hospitals for faster and better management of health data.

Government: Cloud computing has huge significance for the government of a country. Cloud servers are used in several countries to store consolidated information of citizens, which, in turn, helps the government track personal information in terms of taxes, security, and immigration.

Military: Defense forces around the world are heavily reliant on cloud computing technology. For example, in a battlefield scenario, emerging applications help organize and monitor troops relating to remote unmanned areas, health monitoring of soldiers, increasing situational awareness and collecting real-time information, and coordination among geo-distributed troops.

Additionally, cloud applications are widespread, from billing, banking, and shopping to online education and learning, business, and IT enterprises. Cloud computing ensures top-quality customer experience and satisfaction, thereby serving as a key computing technology for the IoT.

1.4 Summary

In this chapter, we briefly presented the history of cloud computing and how it gradually evolved to its current state. The chapter chronologically discussed the cloud deployment models and the different currently available cloud computing services. We further presented the details of cloud-based computations by highlighting the aspects of resource management, virtualization, and green or energy-efficient computation. Finally, we illustrated the popular domain of cloud applications. In the next chapter, we will discuss another key component of IoT – sensor networks and their implications when connected to cloud.

Working Exercises

Multiple Choice Questions

1. Which of the following is not a fundamental cloud service?
   a) Software
   b) Applications
   c) Infrastructure
   d) Platform
2. What is the major disadvantage of the hybrid cloud model?
   a) Compatibility issues while integrating public and private clouds
   b) Complaints issued by legacy organizations against such platforms
   c) Expensiveness
   d) Slow data processing abilities
3. Which of the following are examples of PaaS – (i) Amazon Web Services, (ii) Windows Azure, (iii) Google App Engine, (iv) Cisco WebEx, and (v) Apache Stratos?
   a) (i) and (iii)
   b) (ii), (iii), and (iv)
   c) (ii), (iii), and (v)
   d) (i), (ii), (iii), (iv), and (v)
4. Which of the following is an objective for resource management?
   a) VM placement
   b) Capacity planning
   c) Both (a) and (b)
   d) None of the above
5. The algorithmic approaches towards cloud brokering are:
   a) Binary programming and integral optimization
   b) Mathematical modeling
   c) Genetic algorithm and machine learning
   d) Bayesian optimization

6. Which of the following is not a cloud application – (i) Dropbox (ii) Google Drive (iii) Facebook?
   a) (i) and (iii)
   b) (ii) and (iii)
   c) All of the above
   d) None of the above
7. Green computing focuses on
   a) Reducing CO_2 emission
   b) Reducing environmental impact
   c) All of the above
   d) None of the above
8. What is popularly used for workload modeling?
   a) Linear algebra
   b) Dynamic programming
   c) Binary knapsack
   d) Bag of Tasks (BoT)
9. What is popularly used for VM placement problems?
   a) Machine learning
   b) Dynamic programming
   c) Differential evolution
   d) Particle Swarm Optimization (PSO)
10. What are the parameters measuring environmental impact?
    a) Turnaround time for applications
    b) Application throughput
    c) Cooling costs and power consumption
    d) Price charged by the service provider

Conceptual Questions

1. How can cloud deployment models be classified?
2. What are the main objectives of workload models in resource management?
3. What are the major virtualization models in cloud?
4. In a multiple geo-distributed data center environment, how should the network be assigned within the data centers?
5. How does green computing change the design of cloud applications?

References

[1] P. Mell and T. Grance, "Draft NIST Working Definition of Cloud Computing," Referenced on June 3, 2009. Online: http://csrc.nist.gov/groups/SNS/cloud-computing/index.html.

[2] M. D. Neto, "A Brief History of Cloud Computing," 2014. Online: www.ibm.com/blogs/cloud-computing/2014/03/18/a-brief-history-of-cloud-computing-3/,2014
[3] K. Priya and J. Arokia Renjit, "Data Security and Confidentiality in Public Cloud Storage by Extended QP Protocol," in *Proceedings of International Conference on Computation of Power, Energy Information and Commuincation (ICCPEIC)*, Melmaruvathur, 2017, pp. 235–240.
[4] A. AlQhtani, E. Aloboud, R. Altamimi, and H. Kurdi, "Impacts of VPNs and Firewalls on Public Cloud Performance," *European Modelling Symposium (EMS)*, Manchester, United Kingdom, 2017, pp. 197–201, doi: 10.1109/EMS.2017.42.
[5] H. Wang, "Anonymous Data Sharing Scheme in Public Cloud and Its Application in E-Health Record," *IEEE Access*, 2018, doi: 10.1109/ACCESS.2018.2838095.
[6] Online: https://searchcloudcomputing.techtarget.com/definition/public-cloud, 2009.
[7] Y. J. Hu, "Outsourcing Secured Machine Learning (Ml)-As-A-Service for Causal Impact Analytics in Multi-Tenant Public Cloud," in *Proceedings of the 2nd International Conference on Telecommunication and Networks (TEL-NET)*, 2017, doi: 10.1109/TEL-NET.2017.8343282
[8] Online: https://searchcloudcomputing.techtarget.com/definition/private-cloud [Accessed November 2017]
[9] R. Soares Boaventura, K. Yamanaka, G. Prado Oliveira, E. Rodrigues Pinto, and M. Braz de Andrade Maciel, "Methodology for Statistical Analysis Comparing the Algorithms Performance: Case of Study in Virtual Environments in Private Cloud Computing", *IEEE Latin America Transactions*, vol. 15, no. 2, pp. 333–340, Feb 2017.
[10] D. Cao, P. Liu, W. Cui, Y. Zhong, and B. An, "Cluster as a Service: A Resource Sharing Approach for Private Cloud," *Tsinghua Science and Technology*, vol. 21, no. 6, pp. 610–619, Dec 2016.
[11] R. Birke, A. Podzimek, L. Y. Chen, and E. Smirni, "Virtualization in the Private Cloud: State of the Practice," *IEEE Transactions on Network and Service Management*, vol. 13, no. 3, pp. 608–621, Sept 2016.
[12] R. Jenkins, "Hybrid Public Private Cloud Computing for the Media Industry," *SMPTE Motion Imaging Journal*, vol. 123, no. 3, pp. 56–59, April 2014.
[13] J. M. Wrabetz and E. Weaver, "Leveraging Hybrid Cloud Workflows in Media and Entertainment," *SMPTE Motion Imaging Journal*, vol. 127, no. 4, pp. 16–20, May 2018.
[14] D. Zhu, Y. Guo, J. Chen, Y. Li, and H. Jiang, "Architecture and Security Protection Scheme for Distributed New Energy Public Service Platform with Hybrid Cloud System," *The Journal of Engineering*, vol. 2017, no. 13, pp. 2203–2206, 2017, 10.1049/joe.2017.0721.
[15] J. Zhu, X. Li, R. Ruiz and X. Xu, "Scheduling Stochastic Multi-Stage Jobs to Elastic Hybrid Cloud Resources," *IEEE Transactions on Parallel & Distributed Systems*, vol. 29, no. 06, pp. 1401–1415, 2018.
[16] Y. Zhang, H. Huang, Y. Xiang, L. Y. Zhang, and X. He, "Harnessing the Hybrid Cloud for Secure Big Image Data Service," *IEEE Internet of Things Journal*, vol. 4, no. 5, pp. 1380–1388, Oct 2017.
[17] A. Raja, "Understanding Hybrid Cloud, with Examples," 2016.
[18] Online: www.techopedia.com/definition/26559/community-cloud, 2018

[19] H. Luo, J. Liu, and X. Liu, "A Two-Stage Service Replica Strategy for Business Process Efficiency Optimization in Community Cloud," *Chinese Journal of Electronics*, vol. 26, no. 1, pp. 80–87, 2017.
[20] P. Zhao and X. Yang, "Joint Optimization of Admission Control and Rate Adaptation for Video Sharing over Multirate Wireless Community Cloud," *China Communications*, vol. 13, no. 8, pp. 24–40, Aug 2016.
[21] Y. Wu, M. Su, W. Zheng, K. Hwang, and A. Y. Zomaya, "Associative Big Data Sharing in Community Clouds: The MeePo Approach," *IEEE Cloud Computing*, vol. 2, no. 6, pp. 64–73, Nov–Dec 2015.
[22] I. Petri, J. Diaz-Montes, O. Rana, M. Punceva, I. Rodero, and M. Parashar, "Modelling and Implementing Social Community Clouds," *IEEE Transactions on Services Computing*, vol. 10, no. 3, pp. 410–422, May–June 2017.
[23] Y. F. Purwani and K. Rukun and Krismadinata, "A Review of Cloud Learning Management System (CLMS) Based on Software as a Service (SaaS)," in *Proceedings of International Conference on Electrical Engineering and Informatics (ICELTICs)*, Banda Aceh, 2017, pp. 205–210.
[24] Z. Nikdel, B. Gao, and S. W. Neville, "DockerSim: Full-Stack Simulation of Container-Based Software-as-a-Service (SaaS) Cloud Deployments and Environments," in *Proceedings of IEEE Pacific Rim Conference on Communications, Computers and Signal Processing (PACRIM)*, Victoria, BC, 2017, pp. 1–6.
[25] R. A. Asaka, G. H. S. Mendes, and G. M. D. Ganga, "Factors Influencing Customer Satisfaction in Software as a Service (SaaS): Proposal of a System of Performance Indicators," *IEEE Latin America Transactions*, vol. 15, no. 8, pp. 1536–1541, 2017.
[26] I. Kumara, J. Han, A. Colman, and M. Kapuruge, "Software-Defined Service Networking: Performance Differentiation in Shared Multi-Tenant Cloud Applications", *IEEE Transactions on Services Computing*, vol. 10, no. 1, pp. 9–22, Jan.–Feb 2017.
[27] A. Smirnov, A. Ponomarev, T. Levashova, and N. Shilov, "Platform-as-a-Service for Human-Based Applications: Ontology-Driven Approach," in *Proceedings of the 7th IEEE International Symposium on Cloud and Service Computing (SC2)*, Kanazawa, 2017, pp. 157–162.
[28] C. Dong, Y. Jia, H. Peng, X. Yang, and W. Wen, "A Novel Distribution Service Policy for Crowdsourced Live Streaming in Cloud Platform," *IEEE Transactions on Network and Service Management*, vol. 15, no. 2, pp. 679–692, June 2018.
[29] SynapseIndia – Reviews & Complaints, "Evaluating Risks and Benefits of IaaS, PaaS and SaaS in the Cloud Computing," March 2015.
[30] D. Riane and A. Ettalbi, "A Graph-Based Approach for Composite Infrastructure Service Deployment in Multi-Cloud Environment," in *Proceedings of International Conference on Advanced Communication Technologies and Networking (CommNet)*, Marrakech, 2018, pp. 1–7, doi: 10.1109/COMMNET.2018.8360254.
[31] Y. Chi, X. Li, X. Wang, V. C. M. Leung, and A. Shami, "A Fairness-Aware Pricing Methodology for Revenue Enhancement in Service Cloud Infrastructure," *IEEE Systems Journal*, vol. 11, no. 2, pp. 1006–1017, June 2017.
[32] D. J. Dean, H. Nguyen, P. Wang, X. Gu, A. Sailer, and A. Kochut, "PerfCompass: Online Performance Anomaly Fault Localization and Inference in Infrastructure-as-

a-Service Clouds," *IEEE Transactions on Parallel and Distributed Systems*, vol. 27, no. 6, pp. 1742–1755, June 2016.

[33] M. Guzek, P. Bouvry, and E. Talbi, "A Survey of Evolutionary Computation for Resource Management of Processing in Cloud Computing," *IEEE Computational Intelligence*, Vol. 10, no. 2, pp. 53–67, 2015.

[34] S. Frey, F. Fittkau, and W. Hasselbring, "Search-Based Genetic Optimization for Deployment and Reconfiguration of Software in the Cloud," in *Proceedings of IEEE International Conference on Software Engineering (ICSE)*, 2013.

[35] F. Xu, F. Liu, H. Jin, and A. V. Vasilakos, "Managing Performance Overhead of Virtual Machines in Cloud Computing: A Survey, State of the Art, and Future Directions," *Proceedings of the IEEE*, vol. 102, pp. 11–31, 2013.

[36] G. Jung, K. R. Joshi, M. A. Hiltunen, R. D. Schlichting, and C. Pu, "A Cost-Sensitive Adaptation Engine for Server Consolidation of Multitier Applications," *Middleware*, vol. 5896, pp. 163–183, 2009.

[37] S. Murugesan, "Harnessing Green IT: Principles and Practices", *IEEE IT Professional*, Vol. 10, pp. 24–33, 2008.

[38] A. Qureshi, R. Weber, H. Balakrishnan, J. Guttag, and B. Maggs, "Cutting the Electric Bill for Internet-Scale Systems," in *Proceedings of ACM Special Interest Group on Data Communication (SIGCOMM)*, 2009.

[39] N. Xiong, W. Han, and A. Vandenberg, "Green Cloud Computing Schemes Based on Networks: A Survey," *IET Communications*, Vol. 6 no. 18, pp. 3294-3300, 2011.

[40] R. Nathuji and K. Schwan, "Virtualpower: Coordinated Power Management in Virtualized Enterprise Systems", *ACM SIGOPS Operating Systems Review (OSR)*, Vol. 41, pp. 265–278, 2007.

[41] S. Srikantaiah, A. Kansal, and F. Zhao, "Energy Aware Consolidation for Cloud Computing," *Cluster Computing*, pp. 1–15, Springer 2009.

[42] K. Mistry, C. Allen, C. Auth, B. Beattie, D. Bergstrom, M. Bost, M. Brazier, M. Buehler, A. Capelani, R. Chau, C.-H. Choi, G. Ding, K. Fischer, T. Ghani, R. Grover, W. Han, D. Hanken, M. Hattendorf, J. He, J. Hicks, R. Huessner, D. Ingerly, P. Jain, R. James, L. Jong, S. Joshi, C. Kenyon, K. Kuhn, K. Lee, H. Liu, J. Maiz, B. McIntyre, P. Moon, J. Neirynck, S. Pae, C. Parker, D. Parsons, C. Prasad, L. Pipes, M. Prince, P. Ranade, T. Reynolds, J. Sandford, L. Shifren, J. Sebastian, J. Seiple, D. Simon, S. Sivakumar, P. Smith, C. Thomas, T. Troeger, P. Vandervoorn, S. Williams, K. Zawadzki, "A 45nm Logic Technology with High-k+Metal Gate Transistors, Strained Silicon, 9 Cu Interconnect Layers, 193nm Dry Patterning, and 100% Pb-free Packaging," *IEEE IEDM 2007 Technical Digest*, 10.2, 2007.

[43] M. Chudzik, B. Doris, R. Mo, J. Sleight, E. Cartier, C. Dewan, D. Park, H. Bu, W. Natzle, W. Yan, C. Ouyang, K. Henson, D. Boyd, S. Callegari, R. Carter, D. Casarotto, M. Gribelyuk, M. Hargrove, W. He, Y. Kim, B. Linder, N. Moumen, V. K. Paruchuri, J. Stathis, M. Steen, A. Vayshenker, X. Wang, S. Zafar, T. Ando, R. Iijima, M. Takayanagi, V. Narayanan, R. Wise, Y. Zhang, R. Divakaruni, M.Khare, T. C. Chen, "High-Performance High-

Metal Gates for 45nm CMOS and beyond with Gate-First Processing," *IEEE VLSI 2007 Technical Digest*, 11A-1, 2007.

[44] S. Kemp, "Digital in 2018: Worlds Internet Users Pass the 4 Billion Mark," 2018. Online: https://wearesocial.com/uk/blog/2018/01/global-digital-report-2018 [Accessed on January 30, 2018].

[45] S. Misra, S. Bera, and T. Ojha, "D2P: Distributed Dynamic Pricing Policy in Smart Grid for PHEVs Management," *IEEE Transactions on Parallel and Distributed Systems*, vol. 25, 2014, doi: 10.1109/TPDS.2014.2315195.

Chapter 2

Sensor Networks and the Cloud

The emergence of sensor technology has enhanced the standard of living for mankind, and manifestations of this fact are found in numerous applications, such as robotics [1], vehicles [2], battlefield monitoring [3–5], ubiquitous monitoring [6,7], and target tracking [8–10]. In its early days, sensor nodes were mostly supported by wired communication channels. However, with the advancement of sensor technology, wireless sensors emerged and gained popularity over their wired counterpart because of their portability and easy installation.

A wireless sensor is essentially a device equipped with sensing hardware and a transceiver module. The sensing hardware allows it to sense data from its surroundings; the transceiver module allows the sensor device to receive or transmit data in the form of packets. Thus, every sensor device can be considered a sensor node if it has the ability to sense events and communicate with other such nodes. Hereafter, we use the terms *sensor* and *sensor node* interchangeably. Sensor nodes are usually resource-constrained devices with low power and limited computational abilities. After these sensor nodes were realized in practice and were manufactured at a commercial scale, the concept of Wireless Sensor Networks (WSNs) emerged [11–13]. A WSN comprises multiple sensor nodes that form a network. Every node in a WSN senses through its own hardware and transmits sensed information over the wireless medium to a destination or a sink node via a single- or multi-hop route.

With the heavy demand, usage, and consequent growth of sensor networks, conventional, resource-limited WSNs encountered challenges and difficulties. By 2020, the industrial WSN market is expected to grow by 43.1% annually, and trillions of sensor devices will be Internet connected [7]. This implies that a huge volume of data will be generated and processed by these sensor nodes every day, at every instant. However, conventional WSNs are not sufficiently advanced to handle, manage, and process these huge volumes of data for a number of reasons. Hence, to resolve this problem, cloud computing was adopted as a backbone solution for WSNs.

In this chapter, we present the intrinsic details of WSNs, followed by an illustration of the applicability of cloud computing in WSNs. We also highlight how the integration of the cloud helps us resolve existing issues in data management in WSNs.

2.1 Wireless Sensor Networks

Over the past decades, we have witnessed tremendous growth and evolution of WSNs, which had a huge impact on their various application domains. The sensor nodes in a WSN are deployed over the region, so that one or more physical environmental attributes of the region can be captured and reported. Within these sensor nodes, there are computation and processing modules, which are involved in local processing or pre-processing of the sensed data. However, in terms of computing and storage abilities, sensor nodes are often found to be resource-constrained.

2.1.1 Background and Evolution

The very first sensor node was a temperature sensor developed by Wilhelm von Siemens in 1860. The piezoelectric characteristics and the high resonance stability of the single-crystal quartz [14] paved the way for the advancement of sensor technology for the next decades. The first sensor to be commercialized, in 1883 [13], was a thermostat. It is relevant to note some of the early sensors, including an extremely precise electric thermostat invented by Warren S. Johnson and a Doppler effect-driven motion sensor (used within an alarm system) invented by Samuel Bagno [13].

The definition of sensor was arrived at after a lot of discussion and differences in opinion. Initially, the words “sensor” and “transducer” were used synonymously. The American National Standards Institute (ANSI) definition of a transducer was a device that outputs an electrical signal in response to a measurable quantity. As per the ANSI standard, “transducer” was a more commonly used term. However, based on recent literature, “sensor” has gained more popularity.

2.1.1.1 Historical Survey

The earliest sensor networks were primarily used for defense applications and military purposes. During the Cold War, the Sound Surveillance System (SOSUS) was introduced and used by the United States Army. SOSUS was composed of several acoustic sensors that were strategically deployed to detect and track Soviet submarines. Some of the sensors are still in use by the National Oceanographic and Atmospheric Administration (NOAA) for earthquake monitoring. The operation of these sensor networks was basically hierarchical and took place at different levels.

Around 1980, Defense Advanced Research Projects Agency (DARPA) initiated research on Distributed Sensor Networks (DSNs) [15,16]. One of the co-inventors of TCP/IP protocols and the Director of the Information Processing Techniques Office (IPTO) at DARPA, R. Kahn, took significant initiative in this regard. By then, ARPAnet (the predecessor of the Internet) had already been significantly successful for some years, and Kahn investigated the applicability of the ARPAnet to support low-cost DSNs. The DSNs were assumed to operate autonomously with the possibility of data packets being transmitted to and received by any node. At that time, computers were not as advanced as they are today. Minicomputers like PDP-11 and VAX machines were generally used. The Ethernet was gradually gaining popularity. Technological involvements associated with this project focused on the development of acoustic sensors, communication protocols, self-computing and adaptive algorithms, and distributed software (dynamically modifiable distributed systems and language design). Research at Carnegie Mellon University focused on providing a network operating system for flexible and transparent access to distributed resources needed for a fault-tolerant DSN. At the same time, Accent [15], a communication-oriented operating system, was developed and later evolved to a Mach operating system [17]. Protocols and standards were designed for inter-process communication, dynamic binding, dynamic load balancing, and fault reconfiguration. A test bed was developed using acoustic sensors and Ethernet-connected VAX computers. Researchers at the Massachusetts Institute of Technology (MIT) and the University of Cambridge focused on knowledge-based signal processing techniques for acoustic sensor-based helicopter tracking. The Signal Processing Language and Interactive Computing Environment (SPLICE) was also developed by MIT for efficient data processing and analysis. The MIT Lincoln Laboratory also developed a test bed for tracking aircraft flying at lower elevations using an array of acoustic sensors [18].

Between 1980 and 1990, sensor networks were thought to be composed of a large number of small sensors. However, contemporary technology was unable to support the network architecture. Based on the obtained results from the DARPA-DSN project and the test beds developed, military planners aimed to use sensor networks in network-centric warfare [19]. Prior to network-centric warfare, there was platform-centric warfare, in which sensors were mounted inside weapons within platforms. However, these platforms functioned independently, whereas in network-centric warfare, sensors are not a part of weapons or platforms, but are designed to operate collaboratively, aiding shooters with aggregated information. For example, Cooperative Engagement Capability (CEC) [20] comprised several radars that acquired information about air targets. The information was shared among nodes that had similar interests.

Since then, research on sensor networks has evolved a lot. In the 21st century, we have sensors based on Micro-Electro-Mechanical System and Nanoscale-Electro-Mechanical System technologies, fundamentally wireless sensors with low-power computational abilities. DARPA once again took the initiative of exploring and investigating new technological advancements. To support the technological growth, several communication protocols such as Bluetooth, ZigBee, and WiMax have emerged. Several topological architectures (e.g., centralized, clustered, peer-to-peer, mesh, and hierarchical topologies) have also been developed. The software internal to a sensor is designed to support energy-efficient, autonomous, and latency-sensitive operability, optimized communications, self-adaptability, and security management.

2.1.1.2 Characteristics of WSNs

Several features characterize conventional WSNs:

Hardware: In a WSN, the size of a physical sensor node varies from a few cubic centimeters to few decimeters, and the nodes are deployed according to a topology, which, in turn, is specific to the application being served. A node usually comprises four basic modules [16]: a sensing and actuation module, a communication module, a processing module, and a power module (this excludes other application-specific modules). Each node is designed to contain an open-source operating system that has libraries to support data retrieval and storage, network operations, services, sensing, and other application-specific drivers. The protocols of a sensor node support low-to-moderate data rates and are computationally optimized for minimal energy consumption.

Communication: In order to enable inter-node communication, every node possesses a transceiver that functions within a specific range, called the transmission range of the node. Based on the transmission radio, this range may also be controlled dynamically by the application requirement. A node is designed to communicate over a particular communication standard such as IEEE 802.15.1 (Bluetooth), IEEE 802.15.4 (ZigBee), IEEE 802.16 (WiMax), and so on.

Routing: In a WSN, the sensor nodes generate data packets based on the sensed information. The data packets are then routed to a sink directly or via a multi-hop route using other nodes, based on the topology design of the network. In the early days of WSN, nodes were primitively designed to send data packets directly to a destination. However, this incurred significant energy overhead; several advanced topologies for sensor networks are now prevalent. Based on the topological behavior, the nodes generally route the data packets via a multi-hop route to the sink node. The energy expenditure of a node is attributed to sensing, transmitting, receiving, idle listening, performing internal computations, and storage, among other operations. The routing of information from within the network to the sink is done in a way to minimize the energy consumption of the entire network. Based on topological constraints, there are several schemes for routing within WSNs. Some of the routing schemes are data-centric (data are obtained only from a few specific nodes by querying those nodes); some schemes are hierarchical and comprise clustering and subsequent data aggregation within every cluster, followed by data transmission to the sink node. In addition, routing schemes can be location-based or QoS-oriented.

Internal Software: A sensor operation is supported by open-source operating systems. TinyOS is a popular operating system that enables heavy implementation in code bases with small code size. TinyOS libraries include sensor drivers and tools that enable data collection, storage, and power management. Additionally, the operating system allows maintenance of operational flexibility, which is required due to the unstable nature of the wireless medium.

Data Computation: Computation is a crucial part of sensor networks, as computational strategies have significant impact on the energy efficiency. Computation involves data processing, calibration, fusion, aggregation, and decision making. In-network computation involves meaningful in-network data processing and automatic query handling. Network information processing is also introduced in contemporary WSNs to acquire timely and relevant data from reliable sources. This, in turn, also facilitates dynamic query processing and query channelization within the network.

Network Management: Network management is important in WSNs, as it involves a large number of internal aspects (e.g., reliable data delivery, information routing, efficient utilization of network bandwidth, in-network data aggregation, and dynamic adaptation to topology changes). Therefore, network management focuses on the details of network architecture and communication standards, fault tolerance of the network, resource efficiency, and communication frequencies. Several design models are in use today for network management in WSNs. The layered system structure [21] divides the overall network functionality into distinct layers, thereby achieving modularity. Such systems are easy to handle, as changes can be easily incorporated into the required layer without affecting the others. Another approach focuses on distributing the decision making and control functions into non-overlapping spatially distributed clusters of sensor nodes, making the overall scheme resource-efficient compared to the centralized approach. Policy-based schemes are also popular; nodes communicate based on defined policies, thereby achieving an organized and coordinated network management. Other schemes, such as the information model and service-oriented management, are used in specific cases and topologies of WSNs.

2.1.1.3 WSNs and the Internet

With the increase of WSNs, many applications became reliant on this technology. To support these applications, it was imperative to integrate WSNs into the Internet. Figure 2.1 depicts the evolution of this connection architecture.

Sensor-based communications were no longer intended to be intra-network only. During the 1970s and 1980s [22], the concept of Internet-connected WSNs was developed. The sensor nodes were connected to wired Internet services via Ethernet connections, as shown in Figure 2.1(a). After the invention of cellular networks around 1980s, sensors were connected through wires to mobile-phone towers, as shown in Figure 2.1(b). In rare cases, sensors were equipped with high-power radio transmitters to send packets to phone towers. After about a decade, sensors started transmitting packets using low-power radios on frequency bands of 902 – 928 MHz and 2400 – 2483 MHz, as depicted in Figure 2.1(c). This mode of communication was utilized in the IEEE 802.11 standard. With further advancements in cellular networks, and after the discovery of Wi-Fi, sensors were designed to transmit packets to nearby Wi-Fi routers, from where packets were transmitted to farther destinations over the wireless medium. An example is shown in Figure 2.1(d). By this time, sensor modules were included within mobile phones; for Internet-enabled communication, multiple communication protocols (such as 6LoWPAN) were introduced. Several radio standards, adaptive communication protocols, and technological advancements have enhanced the design of sensor networks. Today, almost every WSN is Internet-connected. As there are too many radio standards, there is also steady market competition to serve WSNs with reduced cost for communication.

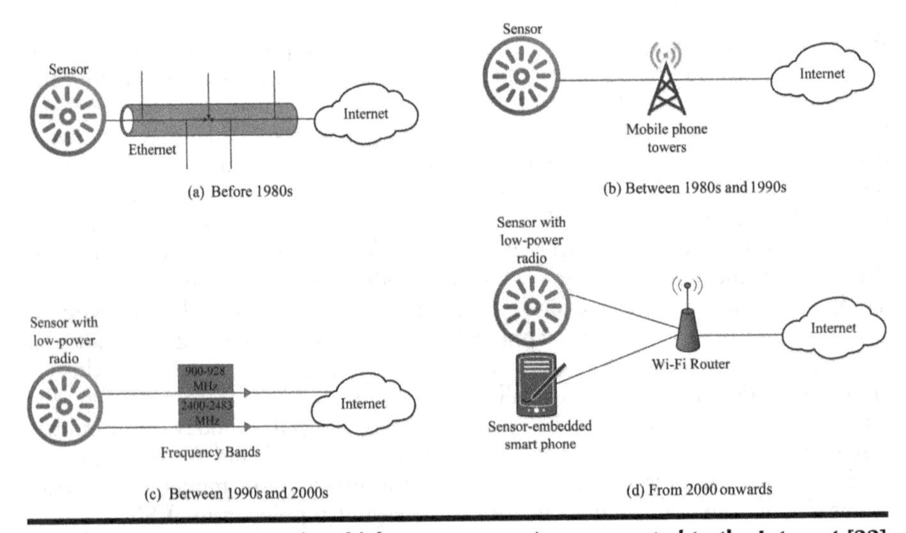

Figure 2.1 Various ways in which sensors were/are connected to the Internet [22]

2.1.1.4 Diversification of Sensor Hardware

The variety of sensor hardware today is huge and widespread. With the widening of the assortment of different types of sensors, a multitude of sensor-based applications has evolved. We briefly present the commonly used sensor types:

1. Infrared sensors are incorporated within specific light sensors to detect light waves in the infrared spectrum. Infrared sensors have several applications (e.g., detecting the presence of an object, measuring the brightness of objects, analyzing patterns, analyzing moisture, determining the absorption feature of gases, and thermal imaging applications).
2. Vision sensors are fascinating camera-equipped sensors that are able to sense external physical attributes of a body inclusive of its dimensions, shape, appearance, and even its location.
3. Fiber optic sensors use optical fibers for their operation and functioning. Some of the sensors are extrinsic and transmit signals from remotely located sensors to a central hub or a sink; intrinsic sensors use fibers to sense different attributes.
4. Level sensors, as indicated by the name, are used to sense and measure the level of any substance. For example, hydrostatic pressure sensors are used to measure the fill level of a liquid. A submersible transmitter is used to measure the pressure due to the weight of liquid above it [16]. Such sensors are also used to monitor water levels for multiple applications or the level of fuel within vehicles.
5. Flow sensors measure the rate of flow of fluids. These sensors are able to detect flows even in very low rates with a high degree of precision. Therefore, a lot of these sensors are used in medical science devices, such as within ventilators, respiratory modules, inhalers, and anesthetic devices [22].
6. Reed switch sensors are especially adept at proximity detection. The sensors are incorporated with reed switches that can be turned on or off using an internal magnet, based on the proximity of external objects or intruders. Reed switch sensors can be widely used for security-based applications, military applications, alarm monitoring, control applications, automated engineering, and robotics.
7. Gas sensors can detect the presence of gas and are hence used for gas leakage detection and safety systems. Such sensors are equipped with a steel exoskeleton [23]. When a gas is close to the sensing element, it gets ionized and is absorbed. This consequently alters the resistance of the element; hence, the amount of current flowing through it also changes.
8. HVAC sensors are essentially control sensors used for heating, ventilation and air conditioning (HVAC) systems. Automated indoor air-conditioning

systems are equipped with HVAC sensors for temperature monitoring and control, dirt and dust control, and humidity monitoring.

9. Motion sensors include passive infrared sensors used for indoor security systems, microwave-based sensors for detecting the reflection of mobile objects, dual technology sensors (combining both infrared and microwave sensors) for mitigating false alarms, area reflective sensors to detect the presence of an object within the threshold proximity of a person or a zone, and ultrasonic sensors to sense and detect reflection of mobile objects [24].
10. Touch-screen sensors can detect touch; when someone touches their touch-responsive surfaces, the current of the electrical signal passing through it is altered, thereby leading to a change in voltage. This change in voltage is used to determine the position of touch.
11. Environment monitoring sensors include sensors for monitoring ambient air temperature, humidity, rainfall, and atmospheric pressure. Data from these sensors are aggregated to obtain meaningful information about local and global weather conditions. The data can be further used to make forecasts and predictions about the future climatic changes.

2.1.2 Design of a Sensor Node

As discussed previously, the basic modules of a sensor node include a sensing and actuation module, a communication module, a processing module, a power module, and other application-specific modules. However, there are other components within a sensor node. In this subsection, we will describe the anatomy of a sensor [14].

In addition to the sensing elements, a node comprises sensor packaging and hardware for signal processing [14]. There is a mechanical or electrical interconnection between sensing element input gates. The node possesses a device for handling the calibration of the sensed data, which means there are inputs, outputs, and actuators for data calibration. There are also signal amplifiers, output gates, and interconnection between the gates. For the transduction of magnetic signals into electrical signals, there is a wide range of choices for selecting the correct sensor material. Some fiber optic magnetic-field sensors, however, use compound sensors for the purpose of transduction.

The sensor technology has evolved immensely. Today, sensors are much more than simple transducers with limited functionalities. Contemporary research focuses on enhancements of sensor transducer mediums, sensor-packaging materials, and the internal design of the devices. Consequently, many advancements are manifested by the synthesis of new transduction materials, microelectronics, and the design and integration of complex systems.

In the last decade, research in sensor technology has reached new heights; it has brought forth "*smart*" sensors. These sensors are equipped with intelligence to deal with the complexities of computation and decision making within themselves, thereby enabling the host systems to enjoy abstraction and simplicity. Therefore, the design of a smart sensor has been enhanced to support advanced functionalities. In addition to the components of a general sensor, a smart sensor includes units for excitation control, analog filtering, compensation, data conversion, and digital communications processing [14]. The primary enabling technology for smart sensors is low-cost microelectronics. On-chip actuators are often used for self-computation. Micro machining techniques have been introduced for building multilayered sensors for increased reliability.

Smart sensor advantages include better maintenance, lower downtime, higher reliability, fault-tolerance, lower weight, and better system architecture. Innovations are still forthcoming in sensor technology and are anticipated to bring tremendous technological progress in the near future.

2.1.3 Applications of Sensor Networks

The applications of sensor networks are many and widespread. In this subsection, we present an overview.

Military Applications: As discussed in Subsection 2.1.1, one of the earliest applications of sensor networks was in the military domain. Sensors were deployed within the bodies of the soldiers to track their movements and their health and to provide surveillance support. Camera-based sensors were used to remotely monitor the activities of the troops and provide dynamic feedback and assistance, accordingly. Sensors were also used inside ammunition dumps to monitor sudden explosions or fire attacks.

Target Tracking: This is one of the most common sensor-based application to date and includes surveillance and intrusion detection applications. These applications generally use motion sensors, camera sensors, accelerometers, and infrared sensors. They are used in surveillance areas, private zones, and secured regions. They can be used to track and detect multiple targets. Tracking mobile targets has a practical significance in vehicular networks. In such cases, mobile WSNs are often used along with static sensors to track and report the mobility of targets.

Healthcare Applications: Healthcare is emerging as a vital application domain for WSNs. Recently, Wireless Body Area Networks (WBANs) have emerged to support remote and ubiquitous healthcare. WBANs comprise several wireless body sensor nodes that sense certain physiological parameters of a human body. These nodes communicate and form a network. Data from this network can be collected and analyzed remotely to diagnose and monitor patients. In recent times, research has significantly advanced in the domain of WBANs. A new wireless communication standard – the IEEE 802.15.6 – has been exclusively introduced for WBAN-based communications. This standard adds to the benefits of data transmission of body sensor nodes by supporting short-range and ultra-low power communication with high data rates.

Household Applications: Sensor networks are useful for several indoor applications. Sensors are deployed within household devices to providing feedback on domestic activities. For example, sensors within refrigerators can periodically report on defrosting conditions and the status of the coolant; microwave ovens can report on the status of the food being heated; air conditioners can be turned on or off based on the presence or absence of the members of the home.

Environmental Applications: Environmental applications are highly relevant in today's world due to global warming and alarming natural conditions. Sensors measuring ambient air temperature, pressure, humidity, rainfall, and soil moisture are very popular. Further, sensors are used to monitor and control pollution. Sensor data from several local zones can be aggregated to draw an inference on the weather conditions of a city or even a state.

Industrial Applications: Sensor networks are used for many industrial purposes. Although sensor-based systems for building automation are expensive in terms of deployment and maintenance, they provide automated control and high precision. Sensor-based industrial applications include quality control, smart robots and structures, environmental monitoring and control within factories, and so on.

Vehicular Applications: Sensor networks are widely used for vehicle monitoring and control. Previously, vehicular systems used Global Positioning System (GPS) to track the location of the vehicles. Today, Global System for Mobile (GSM) technology is also used to determine the name of locations, to find the estimated time of arrival, to detect the status of the driver via alcohol breath sensors, and to track the fuel content of vehicles [5].

Disaster Monitoring Applications: Sensors are used to detect and report seismic activities and volcanic eruptions. For these purposes, sensors for temperature, sound, light, pressure, and acceleration are used [5]. Based on the sensor reports, it is easy to determine the epicenter and the intensity of earthquakes based on sensor readings. It is also possible to determine the area of the zones affected due to volcanic eruptions.

2.1.4 Challenges and Constraints

Many challenges accompany the huge upsurge and enhancement of WSNs, including the following:

1. **Setup**: For any WSN to be operational, sensor nodes must be purchased and deployed over the terrain of interest. Sensor networks are, in general, very expensive and thus are rarely used for personal need. Further, sensors

have to be deployed and the network has established by experienced people. Hence, household customers are not usually the direct beneficiaries of WSNs, as they lack sufficient technical knowledge to operate them.

2. **Topology**: The deployment of nodes varies per the topology of the network, and topological requirements depend on the intended application. If an application changes or if the functionalities of the existing application are modified, the in-network data aggregation algorithm may change accordingly. This, in turn, may call for a topological change. Any change to the setup of a WSN generally requires manual intervention.
3. **Energy**: As mentioned earlier, sensor nodes are extremely resource-constrained and have only a small reservoir of energy during their lifetimes. Once the energy is depleted, it is important to replace or recharge the batteries. When the energy level declines, the node's transmission radius also reduces and directly affects the sensing and computational abilities of the node. Thus, the energy content of the nodes in a WSN is an important metric for choosing strategies related to routing, data aggregation, and other operational protocols. Power conservation is a very serious problem for WSNs.
4. **Maintenance**: Over the lifetime of the network, one or more nodes may experience hardware failures, software errors, or even resource exhaustion. The failure of nodes may rupture the functioning of the network due to issues related to topological breakage, lack of connectivity among nodes, or other routing difficulties. In such cases, it is essential to redeploy or repair failed nodes. However, this is challenging, as redeployment or repairing is primarily manual and can be expensive. Thus, maintenance of WSNs incurs significant overhead.
5. **Hardware**: Every sensor node comes with a vendor-specific proprietary design. Because of monolithic kernels, it is not possible to load and execute multiple customized applications (of the same type) within a single sensor node. However, in some cases, a two-stage boot loader is used. Although such programs help switch and load multiple applications, they involve human intervention and high overhead to enforce the switching. The overhead primarily involves memory management, resource scheduling, load balancing, and process management. Therefore, applications cannot be customized at runtime and executed within traditional WSNs, unless the WSN is reconfigured afresh and the customized application is reloaded within the sensor nodes.
6. **Dissemination**: Existing WSNs are generally single-user centric, meaning only the WSN owner uses the network and its applications. The owner deploys and maintains the network single-handedly. Conventional WSNs do not encourage or support sharing of data and application, especially for security-centric applications such as target tracking, zone monitoring, and terrain

surveillance. Only those organizations that own a network have the right to use it. Therefore, the advantages of WSNs have not been fully realized for private use.

2.2 Unification of WSNs with Cloud: Dawn of a New Era

The history and evolution of cloud computing were covered in Chapter 1. To recap, cloud computing is defined by the National Institute of Standards and Technology (NIST) as an "innovative technological paradigm that provides convenient, on-demand network access to a shared pool of configurable computing resources (for example, sensors) that can be rapidly provisioned and released with minimal management effort or service provider interaction" [25]. Main advantages associated with cloud computing include on-demand resource pooling, huge scalability and elasticity of services, virtualization of resources, abstraction from complex processing logic, and location independence.

Not long ago, cloud computing was considered to be integrated as a key component of conventional WSNs. As mentioned in Section 2.1.4, WSNs have many limitations. The scope and performance of networks are broadened and improved when it unified with cloud computing. The following subsections illustrate the significance and implications of the application of cloud servers in sensor networks.

2.2.1 The Significance of Cloud Computing

It is important to understand why a need was felt to integrate sensor networks and cloud computing. Following are the primary factors that triggered the integration of the two paradigms:

1. ***Insufficient Storage***: Conventional WSNs have simple, server-based architecture. Information from the sensor nodes is dumped into traditional computing servers for storage and analysis. Due to the tremendous growth of sensor networks and the increase in the number of sensor-connected devices, it is almost impossible to maintain a physical server machine for every network or device. Additionally, as the sensors grew "smarter" over time, sensor information became even more complex and voluminous. As cloud computing involves storage within a huge number of gigantic data centers, it was a potential solution to the problem. Cloud data servers were envisioned as storage units for multiple sensor networks and devices, thereby addressing the problem of insufficient storage space. Further, as virtualization is a unique offering of cloud

computing, it is possible to have parallel memory allocations for multiple networks within the same data center.

2. ***Limited Computational Abilities***: Sensor networks require a lot of data processing in and outside of the network. In-network data processing takes place within the sensor node itself. However, nodes possess extremely low computational ability, and excessive computation exhausts the energy of the node. External computations are generally handled within physical servers. As already explained, it is difficult to have distinct servers for each network and to correlate and aggregate data from multiple networks to arrive at a composite inference. Therefore, integrating the paradigm of cloud computing made sense, as cloud data centers offer tremendous computing potential that would prove increasingly useful over time, as the data volume, variety, and velocity increased. Similar to the parallel storage of data from various networks into virtual partitions of the same data center, the processing resources and abilities of data centers can be efficiently allocated and shared among multiple sensor networks.
3. ***Historical Data Analytics***: WSNs collect data from the real world; they analyze and process the data per the need of the application to be served. It might happen that after some time an organization is not interested in the data. Then the question arises with conventional WSNs, "what would happen to all the data collected when the organization would no longer be in need of it"? [26] Due to contemporary technological growth and advancement, data from the real world is expensive and important and cannot be lost or ignored. Thus, the need was felt to store data for historical analysis, investigating future research patterns and trends, and performing later re-analysis. Cloud computing was seen as the platform for storage of old, unused data that are likely to be processed in the future.

2.2.2 How Does Integration Work: Example Scenarios

Researchers have explored the integration of cloud computing from different perspectives. Figure 2.2 depicts the integration of sensor networks into cloud platforms.

To integrate WSNs into cloud platforms, one must obtain access to a public/private/hybrid cloud infrastructure that can be used for storage and processing. The current discussion is on public cloud infrastructure for the sake of simplicity. In this architecture (Figure 2.2), several WSN-based applications can be supported. For any application-specific WSN, the user requests data from the corresponding sensor network. Based on the data or the query request, the data are pulled from the underlying sensor networks and immediately transmitted to

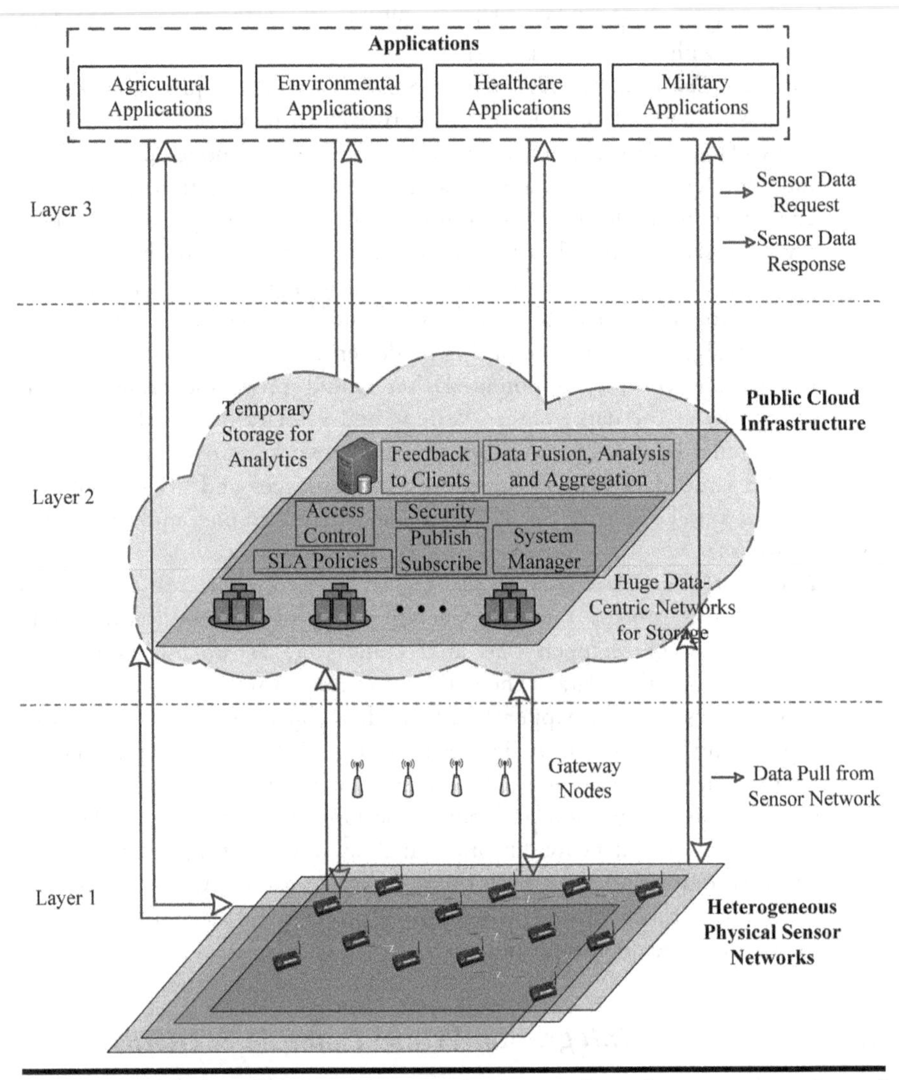

Figure 2.2 A generalized architecture for integration of sensor networks and cloud computing

the cloud end. At the cloud end, the data are stored within the data centers and processed to provide responses to the user queries. Sometimes, data from various data centers must be aggregated and processed before the response is built. In such cases, the data from various sources are dumped to a temporary storage for processing and analysis. Such analyses are dependent on the query type and the storage pattern within the data centers.

2.2.2.1 Data Analytics

Cloud computing provides the platform for data analytics and storage for many WSNs. Following are case studies for integrating cloud computing with sensor networks from the perspective of data analysis.

Mobile Cloud Computing: The framework, as proposed by Zhu *et al.* [27], focuses on the integration of mobile sensor networks to the cloud computing platform. The mobile sensor network is considered clustered, and each cluster of the WSN possesses a sensor gateway. After receiving data packets, the sensor gateway performs data traffic monitoring, data filtering, data prediction, data compression, and data encryption of the packets and forwards them to the cloud gateway. The cloud gateway, in turn, decrypts and decompresses the data and initiates data processing as per the queries obtained from the mobile sensor networks. After data processing terminates, response packets are encrypted, compressed, and transmitted back to the users.

This framework integrates cloud servers with mobile sensor networks to take advantage of the enormous data processing abilities of the cloud. The data-centric architecture of the cloud enables huge and complex data sets to be aggregated and processed in a real-time manner, thereby meeting the requirements of latency-sensitive applications.

Green Mobile Cloud Computing: For mobile cloud computing, sensor data is transmitted from mobile sensor-equipped devices to cloud servers. However, this requires the entire mobile network to be reliable. The green mobile cloud computing framework is low power and designed to address the problems of the resource-constrained mobile sensor-equipped devices.

The cloud servers in this framework are equipped with huge storage capacity, high processing/computation power, scalability, pervasiveness, on-demand self-service, and broad network access, as depicted by De *et al.* [28]. The framework comprises three layers: the sensor nodes, the mobile devices, and the cloud servers. The proposed framework is designed for indoor and outdoor purposes using macrocell and microcell base stations. Results indicate that power consumption can be reduced by 10% and 30%, for indoor and outdoor environments, respectively, if macrocell and microcell base stations are used.

2.2.2.2 Architecture Specific to Applications

Some integration architecture is designed specifically for the application being served, as discussed in the following:

Agriculture: Cloud computing integrates well with agricultural sensor networks to provide services to farmers at very low cost. The agricultural cloud service framework enables farmers to use an inexpensive and interactive interface to request and obtain information from agricultural sensor networks. These sensor networks can monitor water level in the fields, check temperatures, pressure or humidity, monitor rainfall, and so on. Sensors detect and provide real-time information to the cloud servers for storage and processing.

Cloud services include Agricultural Data Acquisition Layer (ADAL), Agricultural-Data Processing Layer (ADPL), and Agricultural Data Storage Service Layer (ADSSL) [29]. ADAL is an example of SaaS that provides multiple interfaces for consumers and focuses on the retrieval of user queries through the Internet. The ADPL, deployed as a PaaS, primarily provides libraries with a uniform format; it processes large data sets to obtain meaningful information. The ADSSL is deployed as IaaS that supports back-end data storage and retrieval.

Using cloud-based agricultural sensor networks, it is possible to process huge volumes of data in real time for disease prediction, image processing of diseased plants, and controlling the cultural environment.

Healthcare: Misra and Chatterjee [7] proposed cloud-assisted Wireless Body Area Network (WBAN) architecture for post-disaster medical relief operations. In this architecture, patients are made to wear body sensor nodes that communicate with the Local Data Processing Unit (LDPU). LDPUs report data to the base station (BS). Without the loss of generality, a BS can be multiple hops away from the LDPU, leading to lengthy and inconvenient packet transmissions from the patient. This gets worse in emergency scenarios when the sensitivity towards transmission latency is extremely low. The solution to the problem proposes an optimized data transmission scheme. Moreover, to organize the endeavor of the medical teams in a methodical manner, the solution integrates the health centers with a health-cloud platform.

Rainforest Rehabilitation: The CSIRO ICT Centre [26,30] deployed sensor networks in the Mt. Springbrook rainforest. The sensor network comprised 640 sensors attached to 185 solar powered nodes for monitoring environmental conditions (e.g., temperature, rainfall, soil moisture, concentration of CO_2, and wind speed). The data collected over long periods is not of immediate use. However, it can be of huge importance in the future for analysis of trends in global warming. Therefore, the data can be presented through cloud services at appropriate charges for usage.

2.2.3 Challenges

The integration of sensor networks to cloud computing has several challenges. Although the integration of sensors and the cloud is very effective in addressing some existing problems with sensor network, some crucial problems need attention. The integration of sensor networks to cloud platforms supports the storage of sensor data within the enormous cloud data centers and enables deep, heavy computation using the cores of the cloud servers.

One intrinsic problem of conventional WSNs is that they are designed in a proprietary, vendor-specific manner. Thus, applications within the sensor nodes are compiled during the manufacturing phase within the monolithic kernels. Hence, a sensor node serves only a particular application during its lifetime. It is difficult to execute multiple applications (of the same type, but in a customized fashion) within a particular sensor node. Therefore, the execution of customized WSN applications is challenging in traditional WSNs.

WSNs are generally single-user centric, thereby preventing other organizations from using them. This results in redundancy and lack of optimized utilization of sensor network resources. Additionally, WSN-based applications require that end-user organizations own the sensor nodes, deploy those over a region, and be responsible for their maintenance. The deployment, maintenance, and management of WSNs are extremely expensive and require the supervision of technical personnel. Laypeople cannot enjoy the emerging sensor technology unless they are directly involved with the purchase, deployment, maintenance, and management of the sensor nodes. The problem is nontrivial, as most end-users are naive and overheads are expensive.

The sensor-cloud is seen as a potential solution to the limitations of conventional WSNs. Sensor-cloud infrastructure acts as the interface between the physical and the cyber world through excellent data scalability, on-demand service provisioning, remote data visualization, and user-programmable analysis. The underlying principle of the sensor-cloud is based on sensor node virtualization in which virtual sensors (VSs) are formed per end-user requirements. Aggregated data from the VSs are transmitted to the end-users in the form of a simple obtainable service (just as electricity or water), named *Sensors-as-a-Service* (*Se-aaS*). Sensor nodes are transformed from typical "hardware" to a simple "service", which is cheap, convenient, user-friendly, and scalable. Details on the sensor-cloud are discussed in Chapter 4.

2.3 Summary

In this chapter, we presented the background and history of sensor networks. The chapter illustrated how the structure and design of WSNs have changed and where they stand today, including the significance of connecting WSNs to the Internet. Conventional WSNs encounter several challenges in real-life scenarios, resulting in the unification of WSNs and cloud computing platforms. The chapter presented the ways in which some existing issues have been resolved through this unification. However, some difficulties remain. Recent research has conceived a potential substitute for conventional WSNs, which are discussed in Chapter 4.

Working Exercises

Multiple Choice Questions

1. The first sensor was commercialized in
   a) 1881
   b) 1882

c) 1883
d) 1884

2. Which of the following is not a valid communication protocol in WSN?
a) IEEE 802.15.1
b) IEEE 802.15.4
c) IEEE 802.16 WiMax
d) IEEE 802.12.1

3. In IEEE 802.11 standard, the frequency bands for communication are:
a) 802–828 MHz and 2200–2383 MHz
b) 902–928 MHz and 2400–2483 MHz
c) 702–728 MHz and 2100–2283 MHz
d) 1002–1028 MHz and 2400–2483 MHz

4. What type of sensor is required for detecting the presence of an object?
a) Level sensor
b) Infrared sensor
c) Fiber-optic sensor
d) None of the above

5. Which of the following are sensor-based applications? (i) environmental applications, (ii) social networking applications, (iii) vehicular applications, (iv) recommendation applications, and (v) military applications
a) (ii) and (iv)
b) (i), (iv), and (v)
c) (ii) and (v)
d) (i), (iii), and (v)

6. Which of the following are the primary advantages of unifying WSNs to cloud? (i) resource elasticity, (ii) virtualization of resources, (iii) reduced rate of hardware failures, (iv) limitations of centralized processing, and (v) on-demand resource pooling
a) (i), (ii), and (v)
b) (ii), (iii), and (iv)
c) (ii) and (iii)
d) (iv) and (v)

7. In a disaster scenario, which of the following may lead to network collapse?
a) Blocked communication channels
b) Higher latency of packet transmission
c) Increased data rate
d) All of the above

8. Which of the following is not a latency-sensitive sensor application?
a) Air pollution monitoring
b) Traffic analysis
c) Surveillance application

   d) Smart parking applications
9. In generalized architecture for the integration of sensor networks and the cloud, which layer contains the applications?
   a) Layer 1
   b) Layer 2
   c) Layer 3
   d) None of the above
10. Which of the following is not an essential component of a sensor node?
   a) Actuator
   b) Transducer
   c) Power module
   d) None of the above

Conceptual Questions

1. Study and investigate the evolution of sensor networks to date.
2. Why was it important to integrate WSNs into the cloud?
3. What are the major goals of sensor integrated cloud platforms?
4. What is the anatomy of a sensor?
5. How did DARPA contribute to the evolution of WSNs?

References

[1] A. Petitti, D. D. Paola, A. Milella, A. Lorusso, R. Colella, G. Attolico, and M. Caccia, "A Network of Stationary Sensors and Mobile Robots for Distributed Ambient Intelligence," *IEEE Intelligent Systems*, vol. 31, 2016, doi: 10.1109/MIS.2016.43.

[2] W. Xiong, X. Hu, and T. Jiang, "Measurement and Characterization of Link Quality for IEEE 802.15.4-Compliant Wireless Sensor Networks in Vehicular Communications," *IEEE Transactions on Industrial Informatics*, vol. 12, no. 5, 2016, doi: 10.1109/TII.2015.2499121.

[3] M. Yunn, D. Bragg, A. Arora, and H. A. Choi "Battle Event Detection Using Sensor Networks and Distributed Query Processing," in *Proceedings of IEEE Conference on Computer Communications Workshops (INFOCOM WKSHPS)*, 2011, doi: 10.1109/INFCOMW.2011.5928912.

[4] N. Watthanawisuth, A. Tuantranont, and T. Kerdcharoen, "Design for the Next Generation of Wireless Sensor Networks in Battlefield Based on Zigbee," in *Proceedings of Defense Science Research Conference and Expo (DSR)*, Aug. 2011, pp. 1–4.

[5] H. Yang, Y. Qin, G. Feng, and H. Ci, "Online Monitoring of Geological CO_2 Storage and Leakage Based on Wireless Sensor Networks," *IEEE Sensors Journal*, vol. 13, no. 2, pp. 556–562, Feb. 2013.

[6] C. Bachmann, M. Ashouei, V. Pop, and M. Vidojkovic, "Low-Power Wireless Sensor Nodes for Ubiquitous Long-Term Biomedical Signal Monitoring," *IEEE Communications Magazine*, vol. 50, no. 1, pp. 20–27, Jan. 2012.

[7] S. Misra and S. Chatterjee, "Social Choice Considerations in Cloud-Assisted WBAN Architecture for Post-Disaster Healthcare: Data Aggregation and Channelization," *Information Sciences*, vol. 284, pp. 95–117, 2014.

[8] C.-C. Hsu, Y.-Y. Chen, C.-F. Chou, and L. Golubchik, "On Design Of Collaborative Mobile Sensor Networks for Deadline-Sensitive Mobile Target Detection," *IEEE Sensors Journal*, vol. 13, no. 8, pp. 2962–2972, May 2013.

[9] S. Misra and S. Singh, "Localized Policy-Based Target Tracking Using Wireless sensor Networks," *ACM Transactions on Sensor Networks*, vol. 8, no. 3, pp. 27.1–27.30, Jul. 2012.

[10] S. Chatterjee and S. Misra, "Target Tracking Using Sensor-Cloud: Sensor-Target Mapping in Presence of Overlapping Coverage," *IEEE Communications Letters*, vol. 18, no. 99, pp. 1435–1438, 2014.

[11] X. Wan, J. Wu, and X. Shen, "Maximal Lifetime Scheduling for Roadside Sensor Networks with Survivability K," *IEEE Transactions on Vehicular Technology*, vol. 64, no. 11, pp. 5300–5313, Nov. 2015, doi: 10.1109/TVT.2014.2381243.

[12] A. Thakkar and K. Kotecha, "Cluster Head Election for Energy and Delay Constraint Applications of Wireless Sensor Network," *IEEE Sensors Journal*, vol. 14, no. 8, pp. 2658–2664, Aug. 2014, doi: 10.1109/JSEN.2014.2312549.

[13] Q. Liang, X. Cheng, and S. W. Samn, "NEW: Network-Enabled Electronic Warfare for Target Recognition," *IEEE Transactions on Aerospace and Electronic Systems*, vol. 46, no. 2, pp. 558–568, Apr. 2010, doi: 10.1109/TAES.2010.5461641.

[14] S. Madria, V. Kumar, and R. Dalvi, "Sensor Cloud: A Cloud of Virtual Sensors," *IEEE Software*, vol. 31, no. 2, pp. 70–77, Mar. 2014.

[15] R. Rashid and G. Robertson, "Accent: A Communication Oriented Network Operating System Kernel," in *Proceedings of the 8th Symposium on Operating System Principles*, vol. 91, no. 8, pp. 64–75, 1981.

[16] K. Sohraby, D. Minoli, and T. Znati, Wireless Sensor Networks: Technology, Protocols, and Applications, Wiley: Wiley-Blackwell, May 2007.

[17] R. Rashid, D. Julin, D. Orr, R. Sanzi, R. Baron, A. Forin, D. Golub, and M. Jones, "Mach: A System Software Kernel," in *Proceedings of the 34th Computer Society International Conference (COMPCON)*, San Francisco, CA, 1989.

[18] R. T. Lacoss, "Distributed Mixed Sensor Aircraft Tracking," in *Proceedings of the American Control Conference*, Minneapolis, MN, 1987.

[19] D. S. Alberts, J. J. Garska, and F. P. Stein, "Network Centric Warfare: Developing and Leveraging Information Superiority," 1999. Online: www.dodccrp.org/NCW/ncw.html.

[20] "The Cooperative Engagement Capability," 1995. Online: http://techdigest.jhuapl.edu/td1604/APLteam.pdf.

[21] M. Yu, H. Mokhtar, and M. Merabti, "A Survey of Network Management Architecture in Wireless Sensor Network," in *Proceedings of the 6th Annual PostGraduate Symposium on The Convergence of Telecommunications, Networking and Broadcasting*, 2006.

[22] L. Burgess. "How Does Sensor Data Go From Device to Cloud?" Worldwatch Institute. Online: http://readwrite.com/2015/10/13/sensor-data-device-to-cloud/.

[23] V. Jain. "Insight – Learn the Working of a Gas Sensor." Online: www.engineersgarage.com/insight/how-gas-sensor-works.

[24] "The Beginner's Guide to Motion Sensors," SafeWise. Online: www.safewise.com/resources/motion-sensor-guide.

[25] "Benefits of Cloud Computing to Wireless Sensor Networks," Jul. 2013. Online: www.wsnmagazine.com/cloud-computing-wsn/.

[26] K. Ahmed and M. Gregory, "Integrating Wireless Sensor Networks with Cloud Computing," in *Proceedings of the 7th International Conference on Mobile Ad-hoc and Sensor Networks*, 2011.

[27] C. Zhu, H. Wang, X. Liu, L. Shu, L. T. Yang, and V. C. M. Leung, "A Novel Sensory Data Processing Framework to Integrate Sensor Networks with Mobile Cloud," *IEEE Systems Journal*, vol. 10, no. 3, pp. 1125–1136, 2016.

[28] D. De, A. Mukherjee, A. Ray, D. G. Roy, and S. Mukherjee, "Architecture of Green Sensor Mobile Cloud Computing," *IET Wireless Sensor Systems*, 2016, doi: 10.1049/iet-wss.2015.0050.

[29] D. S. Mahesh, S. Savita, and D. K. Anvekar, "A Cloud Computing Architecture with Wireless Sensor Networks for Agricultural Applications," *International Journal of Computer Networks and Communications Security*, vol. 2, no. 1, pp. 34–38, 2014.

[30] "CSIRO: Wireless Sensor Network: A New Instrument for Observing Our World," 2010. Online: www.csiro.au/science/Sensors-and-network-technologies.html.

Chapter 3

Introduction to the Internet of Things

The term *Internet of Things* (IoT) was coined by Kevin Ashton of MIT, when he introduced it in a presentation to Proctor and Gamble in 1999 [1]. Since then, the IoT has undergone phases of modulation and modification, which shaped a promising concept into a viable technological solution. The IoT is no longer a technology for the future; it has unfolded as a reality that awaits the correct season to bloom. The emergence of the IoT, as a strong and business-viable solution, was greatly propelled by revolutions in the fields of micro-electro-mechanical systems (MEMS) and wireless communication. IoT devices are designed with sensing, actuating, storage, or processing capabilities. The devices are usually small and are equipped with one or more sensing units, a small amount of memory, and a processing unit. The fundamental idea behind the conceptualization of the IoT is to enable seamless communication among "things" over the Internet. The goal of communication is to allow devices to communicate through message passing and make collaborative decisions that serve one or more pre-determined objectives. The IoT, thus, has found its applicability in a wide range of application domains, from smart healthcare and smart city applications to smart transportation and smart agriculture.

In this chapter, we present the formal definition of the IoT and its intrinsic features. We then highlight implementation aspects of the IoT, including architecture, resource management, and IoT-driven analytics. Lastly, we discuss potential application for the IoT and illustrate the importance of the IoT in these contexts.

3.1 Inception and Background

While the concept of machine-to-machine (M2M) [2] and device-to-device (D2D) [3] communications were budding concepts mostly bound to theoretical and visionary research publications, Kevin Ashton saw the potential of binding these technologies with the things around us and conceptualized a new dimension in ubiquitous computing known as the Internet of "things" [1]. His vision was to revisit and remodel the ways things around us behave and are involved in our everyday life. The objective was to empower things, exploit the principles of M2M and D2D communication, and propose a new network that encompasses the things around us. The two integral parts of the IoT are, hence, sensory and communication. These two are supported by the data analysis and decision making units, which together form the IoT ecosystem. The definition of IoT has undergone several modifications and amendments over time. Following are some of the popular and most accepted definitions of the IoT.

3.1.1 Definition

The primary essence of the IoT is to enable things around us to sense and periodically transmit some environmental parameters. For example, a classical computationally "dumb" refrigerator can be modified to sense certain parameters (e.g., the milk level in the bottle, the number of eggs in the rack, and so on) and become a "smart" *thing*. The transmission of these data from the refrigerator to a smartphone application allows the user to be aware of her shopping list while in the market. Gartner proposed a robust definition for the IoT in *Leading the IoT* [4]:

> The Internet of Things (IoT) is a network of dedicated physical objects (things) that contain embedded technology to communicate and sense or interact with their internal states or the external environment. The connecting of assets, processes and personnel enables the capture of data and events from which a company can learn behavior and usage, react with preventive action, or augment or transform business processes. The IoT is a foundational capability for the creation of a digital business.

Several other organizations proposed definitions for the IoT. Therefore, it was important to standardize the definition to be unambiguous and not have organization-specific attributes or features. Therefore, in 2015, with the vision of standardizing the definition for IoT, the IEEE IoT Initiative published an 86-page document entitled "Towards a Definition of the Internet of Things" to define the IoT through characterization [5]. Instead of proposing a standardized definition for the IoT, in this document, they put forward nine defining characteristics of the IoT: interconnection of things, connection of things to the Internet, uniquely identifiable things, ubiquity, sensing/actuation capability, embedded intelligence, interoperable communication capability, self-configurability, and programmability. Based on these characteristics, we propose the following definition for IoT.

> The IoT can be defined as the interconnection of several uniquely identifiable, ubiquitous, and Internet-connected things that are empowered with sensing/actuation and communication units and are capable of intelligent decision making. The devices are also empowered with interoperability among communication modes and are self-configurable and programmable.

Apart from this, with the evolution of the IoT over the years, several new terms have been proposed by researchers. Some of the most popular terms are *Internet of Everything* (IoE) and *Industrial Internet of Things* (IIoT). IoE is essentially the framework that combines the things with human, data, and processes, offering a richer experience in service provisioning and opening up new areas of innovation and business improvisation [6]. Cisco [7] defines IoE thus:

> The Internet of Everything (IoE) brings together people, process, data, and things to make networked connections more relevant and valuable than ever before, turning information into actions that create new capabilities, richer experiences, and unprecedented economic opportunity for businesses, individuals, and countries.

IIoT is more specific to large-scale industrial IoT. The idea is to establish connectivity among a group of infrastructures and objects to allow for easier management, mining, and access to the data. The infrastructures and objects, however, are expected to require minimal human intervention to generate, exchange, and operate on the data. Boyes *et al.* proposed a comprehensive definition for IIoT [8]:

> A system comprising networked smart objects, cyber-physical assets, associated generic information technologies and optional cloud or edge computing platforms, which enable real-time, intelligent, and autonomous access, collection, analysis, communications, and exchange of process, product and/or service information, within the industrial environment, so as to optimise overall production value. This value may include; improving product or service delivery, boosting productivity, reducing labour costs, reducing energy consumption, and reducing the build-to-order cycle.

3.1.2 Characteristics

The characteristics of IoT may vary based on their application area; however, a few characteristics are intrinsic to the IoT.

(a) *Heterogeneity*: The types of devices that constitute the IoT ecosystem are heterogeneous in regard to aspects such as functionality, size, power consumption, and cost. Each IoT device is designed to sense a set of environmental parameters and report metrics periodically to a hub or sink. The parameters sensed by the IoT devices are widely variable and range from atmospheric temperature, humidity, air pressure, ambient light, and movement to critical physiological parameters such as heart rate, oxygen saturation level in blood, and even ECG and EEG signals. The devices are also heterogeneous in terms of their size and power consumption. While some devices are designed to be highly miniaturized, and can even be implanted within tissues and muscles in a human body, others can be as large as a smart phone or an unmanned aerial vehicle. Based on size, component parts, and functionality, power consumption varies. Most miniaturized and implantable devices are powered by battery and have very low power consumption, thereby allowing them to be functional for long durations, sometimes more than a year. Larger devices with complex functionality often need a constant supply of power or periodic recharging of batteries. In a similar way, driven by the same attributes, device costs vary widely.

(b) *Unique device ID*: Each IoT device is equipped with tags known as unique identifiers (UIDs), which uniquely identify the devices over the Internet. These UIDs are designed subject to the device specifications. One way of implementing the UIDs is to embed a radio-frequency identification (RFID) module within the device. RFID technology is driven by two key components: RFID tags and RFID readers. Miniaturized IoT devices are often equipped with RFID tags, and these tags are read by the RFID readers. Each device is thus identified uniquely. Other modes of unique identification can be propelled by means of communication. A popular technique for doing so is through the Ethernet MAC and IP addresses. When the MAC address is combined with the IP address, each device can be uniquely identified over the Internet [9]. Similarly, phone numbers and SIMs can help in unique identification of the IoT devices.

(c) *Widespread span*: IoT technology is known for its widespread and global span. Because IoT devices or "things" are connected to one another over the Internet, the global connectivity of the Internet allows the IoT to have a wide scope of connectivity and operation. This increased span allows IoT devices to be geo-spatially distributed and to operate in cohesion and cooperation with one another, working towards the same goal or service. This distributed architecture allows the IoT to take the benefits of distributed computing architectures closer to the edge, such as fog computing.

(d) *Interconnected devices*: One of the fundamental features of the IoT is the interconnectivity of devices. In an IoT ecosystem, "things" are expected to operate in cooperation with one another, which together serve a set of goals. For this, communication among the "things" is pivotal. For short-range communication among co-located IoT devices, short-range communication protocols such as Bluetooth, ZigBee, and 6LoWPAN are used [10]. Communication standards are typically designed to facilitate low-power and short-range communication. However, application-specific communication standards serve communication relating to pre-defined traffic types. One popular example of such a standard is the IEEE 802.15.6, which is specifically designed to support ultra low-power, short range (within 10 meters), and high data rate wireless communication in the vicinity of living beings [11, 12]. The applications of this communication standard are mostly foreseen in the field of ubiquitous and remote health monitoring and well-being. However, for devices that are geo-spatially separated and need to communicate, communication takes place over the Internet. IoT devices, in such cases, transmit their messages over wired connections or wireless communication modes, such as Wi-Fi, 4G, and 5G.

(e) *Internet-connected devices*: All IoT devices must be connected with the Internet. Devices that are not geographically co-located and yet need to communicate, do so over the Internet through message-passing. Even devices that are capable of communicating with one another via short-range communication protocols are often constrained in terms of computing and storage resources. Therefore, they must communicate with a stable computing platform, where the data transmitted by the devices are processed and stored. Inter-device connectivity and connectivity with the Internet, therefore, are two key parameters of the IoT.

(f) *Sensing and decision-making capability*: IoT devices can be heterogeneous in terms of their operations. However, a device must possess two specific capabilities to be categorized as a thing, as per the definition of IoT proposed by the IEEE IoT Initiative [5]. The first capability refers to sensing ability. A "thing" must be designed to be capable of sensing one or more environmental parameters and periodically reporting the sensed data. Thus, in most cases, the devices are equipped with one or more sensors. In some cases, the sensors are accompanied by an actuator module.

The second capability refers to decision-making ability. IoT devices are often said to be "smart", as they are empowered with some amount of intelligence to be able to make decisions. These decisions can be made independently or collectively through collaboration with a set of such devices. The intelligence of an IoT device, however, is often subject to the resources available to the device. For example, smartphones, which are equipped with computing and storage resources, are capable of performing complicated in-device decision-making operations compared to a simple ambient temperature-sensing device.

(g) *Ubiquity of services*: IoT devices are often mobile in nature, as are most of the devices requesting the IoT services. Therefore, ubiquity of service provisioning must be offered in IoT-related services. Given the global span of IoT-supporting computing architecture, such as cloud, services are available ubiquitously, irrespective of the location of the requesting device.

(h) *Self-configurability*: Another important characteristic of the IoT ecosystem is self-configurability. Due to the heterogeneity of IoT devices and the vast number of such heterogeneous devices, remote centralized management and controlling is often challenging. Therefore, IoT devices are often designed to be self-managing and self-configurable.

(i) *Programmability*: IoT devices are designed to be programmable. Based on the users' requirements, programmers are able to remotely reprogram devices.

3.2 IoT Middleware

In this section, we present the framework for the IoT middleware. The fundamental principle followed for designing IoT middleware is to minimize human intervention and enhance inter-module interaction so that the device, as a whole, can operate seamlessly and independently. IoT middleware is, therefore, expected to have the following properties: (a) context-awareness, (b) resource management ability, (c) analytical ability, (d) portability, and (e) graphical user interface. We discuss the importance of each property below:

(a) *Context-awareness*: Context – both influenced by the users and dictated by the surroundings – plays an important role in IoT systems. Perera *et al.* classified human-influenced context-awareness into three levels: (i) personalization, (ii) passive, and (iii) active, in their work on context-aware computing in the IoT [13]. When the user explicitly states her preferences concerning demands from the system, then it is classified as personalization. Let us consider the common example of smart lighting and try to explain all three cases. Personalization in this context is when the user explicitly programs the lights in an office corridor to light up if some movement is detected, irrespective of whether it is daytime (with sunlight already lighting up the place) or night. Passive context-awareness is when the movement in the corridor is detected by some IoT devices and suggests the possibility of turning on the lights and may act without the user's intervention. Active context-awareness is when the IoT devices installed in the office can act cooperatively and seamlessly without human intervention. These devices are intelligent enough to detect any movement. Upon detection, they check the luminous intensity of the corridor. The combination of these two parameters dictates whether the lights are required and with what intensity. Similarly, context-awareness of IoT ecosystems based on the surroundings can be classified and defined. One way of achieving this context-awareness is programming the devices in the different ontology modeling languages such as Web Ontology language proposed by the World Wide Web Consortium [13].

(b) *Resource management capability*: IoT devices, often constrained in terms of computing, memory, and storage resources, need to manage and utilize their resources wisely. Particularly, in cases in which a group of such devices with heterogeneous resource specifications operates in cooperation to serve a single goal, efficient distribution of the workload depending on devices' resource specifications is critical. Otherwise, there may be unexpectedly high service latency due to a bottleneck in one device in the group, due to which all others may suffer. An IoT middleware device, thus, needs to be able to solve these problems [14]. One method of efficient resource management for IoT devices is to connect a device to the middleware once before the devices are put into use. During this first connection, the devices actively announce their resource specifications to the middleware platform. The platform is also expected to have information about any new IoT applications. Combining these two knowledge bases, the middleware can decide how to distribute the workload among the different devices.

(c) *Analytical capability*: IoT devices usually generate data in the form of continuous streams, often at high velocity. Such heterogeneous and continuous data streams generated by billions of IoT devices constitute a huge volume of data, often referred to as big data. Analyzing this volume of data to extract meaningful information is one of the key requirements of IoT systems. Given the resource-constrained nature of IoT devices, processing and analysis of the data can be made easier and closer to the devices, if the IoT middleware platform can be empowered with some computing and storage resources. This would make processing or pre-processing the data streams faster and might even reduce service latency. The heterogeneity of the incoming data streams also makes the data processing and storage harder. One way to store/process data is to use NoSQL databases [15]. IoT middleware, therefore, is expected to provide support for NoSQL databases.

(d) *Portability*: Portability is a key requirement for IoT to thrive as technology for the future and have a powerful business potential. IoT devices, therefore, must be designed to facilitate plug-and-play. This will make the technology ubiquitous and will become a universal technology unrestricted by compatibility issues and lack of portability. IoT middleware platforms are expected to play a critical role in this aspect of portability and seamless operability. However, with an increase in portability come security concerns. Allowing devices to connect seamlessly makes them more vulnerable to platform-specific attacks. As the backbone of the IoT ecosystem, IoT middleware is expected to provide additional security guarantees. da Cruz *et al.* [16] proposed the essential security guarantees IoT middleware must offer: device authentication, authentic data publishing, access control, and device identification.

(e) *Graphical user interface*: From an end-user's perspective, it is important that the platform be equipped with a simple graphical interface. In a smart car, accompanied by several smart sensors (i.e., IoT devices), this graphical user interface (GUI) can be the car-dashboard, where the technologically naive end-user can get a visual projection of the operation of IoT sensors. However, a big gap needs to be filled because, due to the lack of native GUIs, we have to integrate third-party applications, such as Freeboard [17] or Grafana [18] to provide dashboard-support [16].

3.3 IoT Applications

In today's data-driven business model, the IoT has found its applicability in almost all major fields of modern well-being facilities, production industries, and smart technology's design for the future. The devices in an IoT system often have different purposes, data rates, and sizes. This allows a large number of devices to be a part of IoT, constituting a vast and widespread network of devices. The IoT is found in several fields, such as healthcare, agriculture, manufacturing industries, supply-chain management. In this section, we present the impact of the IoT in a few major application areas.

3.3.1 Healthcare

Healthcare is one of the most important application domains and has a number of component use cases, each with a different set of specifications. For example, the requirements of in-hospital care are significantly different from those of remote and ubiquitous health monitoring, as the latter involves a high degree of patient mobility [19, 20]. The IoT can benefit most of these use cases through its diverse service model. A patient is connected to a number of sensor-equipped devices, each of which monitors a particular physiological parameter (e.g., body temperature, heart rate, and glucose level in blood). The devices are uniquely identifiable through RFID tags or MAC addresses. Data sensed by these devices are sent to a sink node, where all data streams are pre-processed; the devices may even communicate with one another through the sink. The pre-processed data are periodically sent to the computing hub such as cloud, for real-time analysis and long-term data storage. Support of a computing framework, such as the cloud, also ensures that, even if the patients are mobile, ubiquitous data management and service provisioning can be maintained [21, 22]. Alongside healthcare, these wearable devices are used in less critical application scenarios, such as well-being and monitoring of athletes.

3.3.2 Smart City Applications

The use of IoT is probably most extensive in the context of smart city applications. Most modern vehicles are equipped with a large number of sensors. A typical car, has several varieties of sensors, including mass air flow, engine speed, anti-lock brake, oxygen, and fuel temperature sensors. These sensors together transform a car into a smart "thing". On the roads, we have traffic monitoring sensors and cameras, parking lot monitoring sensors, and pedestrian detention sensors to facilitate smart traffic management. Clearly, inter-module and inter-device communication are crucial to robust and efficient traffic management system. The different defining properties of the IoT allow us to realize these requirements of the system in practice. Other similar smart city applications include smart water distribution, smart power distribution and management, and environment monitoring.

3.3.3 Smart Home

Another closely related and rich use case of the IoT is the smart home. The smart home is one of the fastest growing IoT application domains, with more than 60,000 users moving towards smarter homes. This is supported by the several companies that provide IoT-based smart home solutions. Smart locks, smart sprinklers, and smart cooling/heating are the most popular smart home applications. In most cases, the devices are sensor-equipped and centrally controlled and managed by a home-specific aggregator module.

3.3.4 Telecommunication

The IoT has the potential to merge and unify diverse telecommunication technologies. One common example can be observed in modern SIM cards, where communication technologies, such as near field communication (NFC), Bluetooth, GSM, and WLAN are combined and put into a single module, which is also supported by a multitude of sensor modules [23]. Data sensed by these sensors are communicated over one of the many communication means. The devices, thus, can communicate among themselves. Another use of the IoT in enhancing the communication framework is the constitution of large-scale peer-to -peer (P2P) networks of smart "things". These things are capable of communicating with one another and contributing towards one or more objectives.

3.3.5 Supply Chain Management

Applications of IoT technology in supply chain management offers multiple benefits. Let us take the example of car manufacturing and delivery system. In assembly, the car's components are fitted with RFID-tags. As these components become easily accessible and identifiable with RFID readers, tracking of the components is very easy, even in a large-scale manufacturing and assembling plant. This also facilitates automated and unmanned inventory management and tracking out-of-stock components and misplaced or stolen items. Once assembled, when the cars are ready to be delivered, a GPS module planted on the transportation vehicle and on the cars can help in remote monitoring of the delivery procedure. IoT, thus, is an important part of most production-based and assembly-based industries.

IoT also has the potential to facilitate smart agriculture and smart farming. Unmanned crop health monitoring, rainfall measurement, and temperature measurement are few of the many applications that can benefit from the deployment of large-scale agricultural sensor nodes that operate in synchrony and coherence with a central decision-making module. Other application domains include smart retail, smart living, industrial Internet, media and entertainment, and smart object recycling.

3.4 Open Research Challenges and Future Trends

The IoT is highly reliant on a number of component technologies. Each of the multiple technological facets of the IoT has a specific set of requirements, which, in turn, are met by new disruptive technologies. Although the IoT is already put into practice, a number of challenges must be overcome before we can have a city-scale, fully operational, coherent IoT-driven ecosystem.

For example, ubiquitous and seamless *sensing* is a fundamental part of the IoT technology, and sensor-cloud is a new sensing paradigm, which offers sensing-as-a-service, to enable this part of the IoT. Similarly, low-latency and location-specific *computing* are key requirements for most IoT services. To facilitate this, we leverage the benefits of the new computing paradigm, fog computing, which extends the cloud's services closer to the edge of the network and provides IoT services in real time. These are the pillars of the IoT. We discuss the influence and significance of these two technologies on IoT, but a number of research and technological challenges must be addressed to establish a global-scale seamlessly functional IoT ecosystem.

(a) *Identity management*: The founding blocks of the IoT are the billions of Internet-connected and uniquely identifiable devices. The challenge here is to manage the huge number of devices uniquely over the network. The fast growth in the device count has prompted the introduction of IPv6 addressing. With its huge and apparently never-ending address-space, IPv6 is capable of supporting unique addressing for all IoT devices. However, mapping the sensor-addresses with the IPv6 addressing scheme remains a challenge. This also calls for the need to build services to focus on the local and global address resolution.

(b) *Managing heterogeneity*: Heterogeneity is an integral part of the IoT. With different devices operating on different principles and communicating through different communication standards (such as Bluetooth, ZigBee, and Wi-Fi), the research challenge is to unify all devices under a common umbrella. In this direction, in recent times, work has been done on standardization of IoT protocols, particularly for device discovery, inter-device communication, and sensor data management.

(c) *Energy management*: For most IoT devices, resource is a major constraint. Offloading the computing and storage tasks to the supporting computing framework effectively compensates for the lack of computing and storage resources of the IoT devices. The major bottleneck that remains, however, is the scant availability of energy. As most IoT devices are powered by small batteries with finite energy resource, careful management of this energy is crucial. Regulating the sleep-cycle of the devices, the frame transmission policy, and the computational load of the IoT devices is thus one of the critical challenges in this regard.

(d) *Service orchestration*: The requirement for latency-sensitive real-time services triggered the introduction of the fog computing tier and a close-to-the-edge computing paradigm [24, 25]. While theoretically fog can provide location-specific and real-time IoT services, the challenge is to orchestrate the services between the fog and the cloud tiers. Decisions concerning which IoT services are to be provided to which fog tier, and which are to be redirected to the cloud, are still difficult to make.

(e) *Load balancing*: Efficient load balancing is another important challenge for the IoT that helps in extending the lifetime of devices. One way to balance this load for IoT workload is to use the process of clustering. In clustering, the entire IoT network and devices are partitioned into clusters or groups. Each cluster is led by a cluster head, and these heads coordinate with one another for overall communication. The advantages of clustering include reduction of crucial parameters such as network bandwidth, energy consumption, and emission of redundant data packets, while increasing the network lifetime.

(f) *Application design*: The IoT encompasses a wide range of applications, each subject to a specific set of requirements. Several challenges are intrinsic to facilitating uninterrupted operation of these applications. One of the biggest challenges is determining how the IoT ecosystem should function when humans become part of the operational loop. Also, as application requirements constantly change, a seamless, automated, and dynamic service association with the applications is a big technological challenge.

(g) *Programming models*: Owing to the diversity of the application requirements and the application-hosting platform, it is very challenging to design a standard programming model for the IoT. A distributed programming approach is more applicable for services that function in the fog computing tier. On the other hand, for the cloud data centers, a centralized model is followed. In this context, the concept of reactive programming has gained popularity among IoT developers and programmers.

(h) *Security and privacy*: The scale and heterogeneity of IoT devices and services pose multiple security and privacy related challenges. The security challenges encountered in an IoT ecosystem are primarily targeted towards securing the IoT architecture and protecting IoT devices from malicious attacks. In the first case, the security measures are expected to be embedded into the architecture. The latter case requires detection and avoidance of malicious denial-of-service and distributed denial-of-service attacks originating from outside the IoT network. As far as privacy is concerned, the IoT deals with a large amount of personal data, which are processed by the service providers. Withholding the privacy of this personal data is one of the biggest challenges in a multi-stakeholder distributed IoT ecosystem. Moreover, with the enforcement of the General Data Protection Directive in the European landscape, protection of privacy of personal data of the citizens is more critical an issue than ever [26].

3.5 Summary

In this chapter, we have familiarized readers with the concept of the IoT. We discussed its inception and definition. We described IoT middleware and elaborated on different application domains of the IoT, with examples. Finally, we discussed the open research challenges that need to be addressed and possible future research directions.

Working Exercises

Multiple Choice Questions

1. Which one of the following is not a defining characteristic of the IoT as per the IEEE IoT Initiative documentation?
   a) device interconnectivity
   b) device heterogeneity
   c) connectivity with the Internet
   d) locality-specific services
2. Which one of the following is not a necessary feature of IoT middleware?
   a) portability
   b) light-weight
   c) context-awareness
   d) supporting GUI
3. Which one of the following is not a communication technology for short-range communication in the IoT?
   a) Bluetooth
   b) IEEE 802.15.6
   c) ZigBee
   d) none of the above
4. Which means is/are used for unique identification of IoT devices?
   1) MAC address
   2) RFID tags
   3) both (a) and (b)
   4) none of the above
5. Which of the following are potential IoT threats?
   a) sensor type
   b) security
   c) device heterogeneity
   d) none of the above

Conceptual Questions

1. What are the key defining characteristics of the IoT?

2. Discuss how you would design middleware that would be suitable for the IoT.
3. What are the applications of the IoT in healthcare? Can you imagine how the IoT can benefit remote ambulatory healthcare?
4. Describe your views on how the IoT can influence a smart city environment by illustrating three use cases.
5. Mention five key research and technological challenges faced during the large-scale seamless operation of the IoT.

References

[1] K. Ashton, "That 'Internet of Things' Thing," *RFID Journal*, 2009. Online: www.rfidjournal.com/articles/pdf?4986.

[2] G. Wu, S. Talwar, K. Johnsson, N. Himayat, and K. D. Johnson, "M2M: From Mobile to Embedded Internet," *IEEE Communications Magazine*, vol. 49, no. 4, pp. 36–43, 2011.

[3] A. Asadi, Q. Wang, and V. Mancuso, "A Survey on Device-to-Device Communication in Cellular Networks," *IEEE Communications Surveys & Tutorials*, vol. 16, no. 4, pp. 1801–1819, 2014.

[4] Gartner, "Leading the IoT," ed. M. Hung, 2017. Online: https://tinyurl.com/ycq37ovs [Accessed: April 2019].

[5] A. B. Chebudie, R. Minerva, and D. Rotondi, "Towards a Definition of the Internet of Things (IoT)," *IEEE Internet Initiative*, 2015. Online: https://tinyurl.com/yckrfolc [Accessed: April 2019].

[6] M. H. Miraz, M. Ali, P. S. Excell, and R. Picking, "A Review on Internet of Things (IoT), Internet of Everything (IoE) and Internet of Nano Things (IoNT)," *Internet Technologies and Applications (ITA)*, pp. 219–224, 2015.

[7] "Internet of Everything," Cisco. Online: https://newsroom.cisco.com/ioe [Accessed April 2019].

[8] H. Boyes, B. Hallaq, J. Cunningham, and T. Watson, "The Industrial Internet of Things (IIoT): An Analysis Framework," *Computers in Industry*, vol. 101, pp. 1–12, 2018.

[9] "Identifiers in Internet of Things (Iot) Version 1.0," Alliance for Internet of Things Innovation (AIOTI), *WG03 – IoT Standardisation*. Online: https://tinyurl.com/ya4d4fhs, 2018, [Accessed April 2019].

[10] M. Elkhodr, S. Shahrestani, and H. Cheung, "Emerging Wireless Technologies in the Internet of Things: A Comparative Study," *International Journal of Wireless & Mobile Networks*, vol. 8, no. 5, pp. 67–82, 2016.

[11] S. Sarkar, S. Misra, B. Bandyopadhyay, C. Chakraborty, and M. S. Obaidat, "Performance Analysis of IEEE 802.15. 6 MAC Protocol under Non-Ideal Channel Conditions and Saturated Traffic Regime," *IEEE Transactions of Computers*, vol. 64, no. 10, pp. 2912–2925, 2015.

[12] S. Sarkar, S. Misra, C. Chakraborty, and M. S. Obaidat, "Analysis of Reliability and Throughput under Saturation Condition of IEEE 802.15.6 CSMA/CA for Wireless

Body Area Networks," in *Proceedings of IEEE Global Communications Conference (GLOBECOM)*, 2014, pp. 2405–2410.

[13] C. Perera, A. Zaslavsky, P. Christen, and D. Georgakopoulos, "Context Aware Computing for the Internet of Things: A Survey," *IEEE Communications Surveys & Tutorials*, vol. 16, no. 1, pp. 414–454, 2014.

[14] M. A. Chaqfeh and N. Mohamed, "Challenges in Middleware Solutions for the Internet of Things," in *Proceedings of International Conference on Collaboration Technology Systems*, 2012, p. 2126.

[15] P. Paethong, M. Sato, and M. Namiki, "Low-Power Distributed NoSQL Database for IoT Middleware," in *Proceedings of the 5th ICT International Student Project Conference*, 2016, pp. 158–161.

[16] M. A. A. Da Cruz, J. J. P. C. Rodrigues, J. Al-Muhtadi, V. V. Korotaev, and V. H. C. de Al-Buquerque, "A Reference Model for Internet of Things Middleware," *IEEE Internet of Things Journal*, vol. 5, no. 2, pp. 871–883, 2018.

[17] "Freeboard – Dashboards for the Internet of Things." Online: https://freeboard.io/ [Accessed April 2019].

[18] "Grafana – The Open Platform for Analytics and Monitoring." Online: https://grafana.com/ [Accessed April 2019].

[19] S. Misra and S. Sarkar, "Priority-based time-slot allocation in wireless body area networks during medical emergency situations: an evolutionary game theoretic perspective," *IEEE Journal of Biomedical and Health Informatics*, vol. 19, no. 2, pp. 541–548, 2015.

[20] S. Sarkar and S. Misra, "From Micro to Nano: The Evolution of Wireless Sensor-Based Health Care," *IEEE Pulse*, vol. 7, no. 1, pp. 21–25, 2016.

[21] S. Sarkar, S. Chatterjee, S. Misra, and R. Kudupudi, "Privacy-aware blind cloud framework for advanced healthcare," *IEEE Communications Letters*, vol. 21, no. 11, pp. 2492–2495, 2017.

[22] S. Sarkar, S. Chatterjee, and S. Misra, "Evacuation and Emergency Management Using a Federated Cloud," *IEEE Cloud Computing*, vol. 1, no. 4, pp. 68–76, 2014.

[23] D. Bandyopadhyay and J. Sen, "Internet of Things: Applications and Challenges in Technology and Standardization," *Journal of Wireless Personal Communications*, Springer, vol. 58, no. 49, pp. 1–24, 2011.

[24] S. Sarkar, S. Chatterjee, and S. Misra, "Assessment of the Suitability of Fog Computing in the Context of Internet of Things," *IEEE Transactions on Cloud Computing*, vol. 6, no. 1, pp. 46–59, 2018.

[25] S. Sarkar and S. Misra, "Theoretical modelling of fog computing: A green computing paradigm to support IoT applications," *IET Networks*, vol. 5, no. 2, pp. 23–29, 2016.

[26] Subhadeep Sarkar, Jean-Pierre Banâtre, Louis Rilling, and Christine Morin, "Towards Enforcement of the EU GDPR: Enabling Data Erasure," in *Proceedings of IEEE International Conference of Internet of Things*, pp. 1–8, 2018.

THE SENSOR-CLOUD PARADIGM

Chapter 4

The Sensor-Cloud vs. Sensors and the Cloud

In Chapter 2, we observed the limitations of conventional WSNs in terms of maintenance, deployment, resource consumption, and performance. We illustrated the benefits of the integration of cloud computing to sensor networks for addressing and resolving several existing issues related to the performance of WSNs [1,2,24]. We also realized that although the integration of sensor networks to cloud platforms was hugely constructive and prosperous, multiple limitations and difficulties were left unsettled and unaddressed.

This chapter presents and discusses a very recent and advanced computing paradigm: sensor-cloud, which was conceived of as a potential substitute for conventional WSNs. The sensor-cloud paradigm is distinctly different from the integration of sensors and the cloud and should not be considered a mere integration of the two platforms. This chapter will present the principles of the sensor-cloud paradigm and the advantages it has to offer. Before introducing and illustrating the details of the sensor-cloud, we briefly highlight the major limitations of cloud-assisted WSNs.

One crucial problem that persists is that, with the state-of-the-art (i.e., traditional) WSNs, most people cannot enjoy the emerging sensor technology without being directly involved with the purchase, deployment, maintenance, and management of the sensor nodes. The problem is practical and the solution non-trivial, as most end-users are naive and the overhead associated with WSNs is expensive. Existing WSN-based applications require the user-organizations to own the sensor nodes, deploy those over the terrain, and be responsible for their maintenance. As mentioned, WSNs are mostly single-user centric; hence, the renderability of customized applications in conventional WSNs is almost infeasible. Organizations that do not own sensor nodes are deprived of access to applications of WSNs deployed in the field. This applies to any WSN-specific application and results in the redundancy and unoptimized utilization of sensor-network resources. Further, deployment, maintenance, and management of WSNs are very expensive and require the supervision of technical personnel.

4.1 What Is the Sensor-Cloud?

Recent research (e.g., [3,4]) have vastly considered sensor-cloud infrastructure as a potential substitute for traditional WSNs. Sensor-cloud infrastructure, which is essentially an extension of conventional cloud computing [5–8], thrives on the principle of virtualization of physical sensor nodes, thereby rendering a powerful infrastructure that interfaces between the physical and cyber worlds. According to MicroStrain[1], one of the pioneers in this technology, sensor-cloud infrastructure is defined as [3]:

> A unique sensor data storage, visualization and remote management platform that leverage powerful cloud computing technologies to provide excellent data scalability, rapid visualization, and user programmable analysis.

The principal benefit of the sensor-cloud is that it disseminates use of physical sensors to end-users who do not own, deploy, or manage the physical sensor nodes [9]. In order to use the sensor-cloud infrastructure, end-users are required to possess their own sensor-based applications. These applications are fed by the sensed information from the sensor networks by the sensor-cloud service provider on-demand. The underlying procedures of obtaining the raw sensed data and processing and aggregating it into meaningful information are completely abstracted to the end-users. The physical sensors are virtualized to form Virtual Sensors (VSs), and data from the VSs are sent to the end-users. From the end-user organization perspective, it appears that the organization is continuously served by a dedicated sensor node. However, in reality, a particular physical sensor node serves multiple end-users and is dynamically allocated or deallocated to serve different end-users. Thus, the virtualization of the physical sensor nodes enables the end-users to envision the sensors in the form of a service, commonly known as *Sensors-as-a-Service* (*Se-aaS*). Unlike traditional WSNs, Se-aaS creates a new perception and breaks the conventional way of envisioning the sensor nodes, not just as typical hardware, but as simple obtainable service, just like water or electricity [10–12].

4.2 Background of the Sensor-Cloud

In this section, the background concepts and the architectural details of sensor-cloud infrastructure are illustrated, including various actors and their interrelationships, functional components of the infrastructure, and the workflow of the platform.

4.2.1 Motivation of the Sensor-Cloud

The sensor-cloud is essentially a cloud-based platform for heterogeneous sensor data storage, processing, visualization, and management. The primary idea behind the conceptualization of this infrastructure is seamless service provisioning of sensors through virtualization [13], in which the users are completely unaware of the physical location of the sensor nodes. Virtualization creates a complete abstraction of the underlying physical sensor nodes, independent of the network topology. Figure 4.1(a) depicts a traditional WSN in which the user is the sole owner of the network. The user is the only one authorized to utilize the data obtained from the network. Figure 4.1(b) shows the same topology of sensor nodes based on virtualized architecture. Thus, although the underlying topology is identical, the network serves multiple end-users with varying sensor data

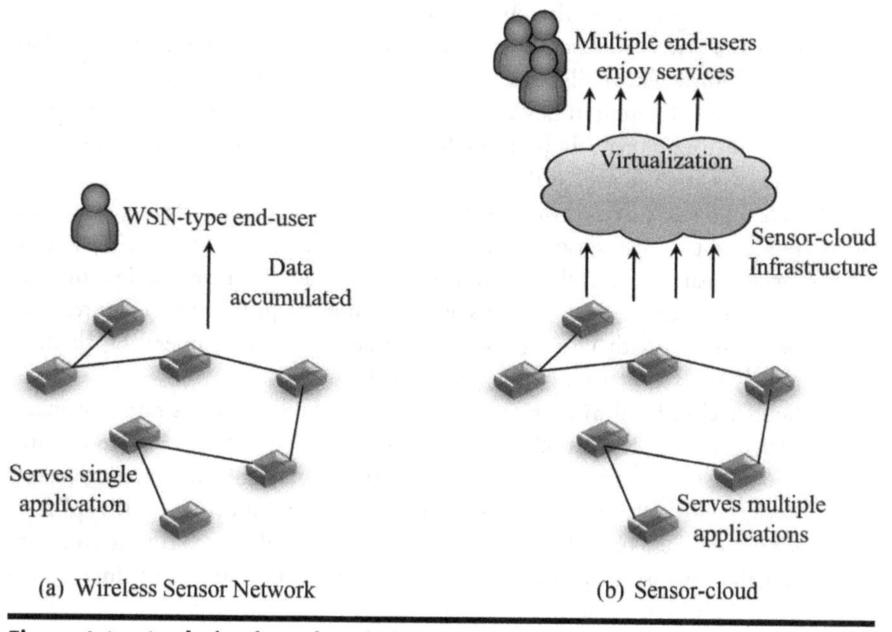

Figure 4.1 Analysis of topology independence [10]

requirements. The architecture of the sensor-cloud is independent of the underlying network topology or the orientation of sensor resources. The success of applications in a sensor-cloud environment is the wide dissemination of the overall sensor network technology to every user (with an application, but not necessarily with technical knowledge and experience), even if s/he does not own sensor nodes. This reduces the responsibilities and overhead associated with maintenance, replacement, re-deployment, and hardware management. Thus, every user-organization envisions Se-aaS, instead of visualizing sensors as conventional physical hardware devices.

From the perspective of cost-effectiveness, Se-aaS is more economic than traditional WSNs [6], mainly because a user-organization is relieved from the initial deployment costs and the auxiliary management costs for repairing and reinstallation. The services of Se-aaS for every user are quantified into measurable units, and user-organizations are charged for the consumable units only. This pay-as-you-go model leads to a fair pricing policy.

4.2.2 Actors of the Sensor-Cloud

Sensor-cloud infrastructure comprises three different actors [7,14]:

1. **End-user**: An end-user is a person (or an organization) who (that) possesses his/her (its) own applications that are to be fed with sensor data from the physical sensor networks. As the type and amount of the demand changes with time, the end-users enjoy scalability of Se-aaS, provided by the cloud service provider (CSP) (i.e., the end-users can demand different sensor services at different times from heterogeneous sensor devices; services are offered instantaneously by the CSP). In return, end-users pay according to their usage of Se-aaS to the CSP.
2. **Sensor-owner**: Sensor-owners are business actors of the sensor-cloud. They purchase physical sensor devices and lend these devices to the CSP. Sensor-owners earn a monthly monitory profit based on the usage of their respective sensor devices. This is analogous to people who register their personal cars with renting companies that take care of operating them on a daily basis and pay a rental profit to the owners.
3. **Sensor-cloud administrator**: The sensor-cloud administrator primarily manages and controls all data processing activities within the cloud infrastructure (e.g., activities related to virtualization of the physical sensor devices into distinct VSs, maintenance and monitoring of the physical sensor devices, organization of the unstructured data, executing computationally intensive queries over the big data sets, and real-time service provisioning of Se-aaS).

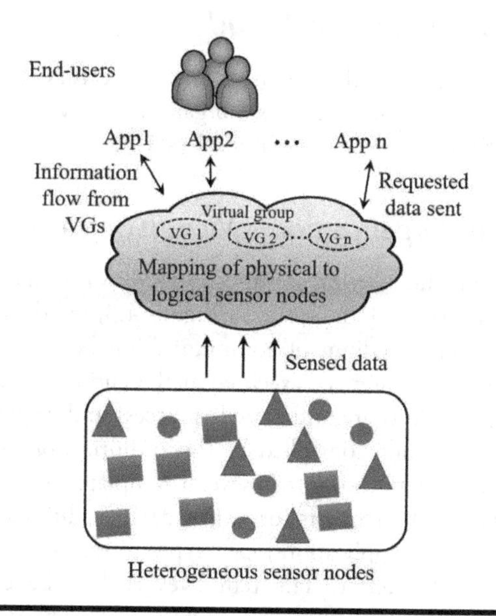

Figure 4.2 Architecture of the sensor-cloud [10]

4.2.3 Architecture of the Sensor-Cloud

The fundamental difference between conventional WSNs and sensor-cloud platforms is that in the sensor-cloud platforms, applications are decoupled from the physical sensor nodes. The sensor nodes are used only for sensing and performing local data aggregation, while all application-processing logic is handled within the cloud-computing unit.

Sensor-cloud infrastructure essentially follows a three-tier architecture [7], as shown in Figure 4.2. Tier 1 is the bottom layer and contains physical sensor nodes that are heterogeneous in terms of vendors, sensing hardware and sensing capabilities, communication modules and mechanisms, and applications. The data from tier 1 are pulled by the cloud-computing unit that comprises tier 2. All computations are performed in tier 2 (i.e., tier 2 is the center for execution of sensor allocation and deallocation algorithms, sensor grouping and virtualization, pulling sensor data from tier 1, aggregation of raw sensor data, and transmission of aggregated sensor data to tier 3). Tier 3 contains multiple end-user organizations with their own applications. Data from tier 2 are fed to these applications in runtime; hence, modifying the structure and design of the applications is an independent operation that does not call for any change in tier 1 for fresh application reloading or reconfiguration of sensor nodes.

4.2.4 Views of the Sensor-Cloud

In this subsection, we present the details of the architectural aspects of the sensor-cloud from two perspectives: (a) the user-organization's view or the logical view, and (b) the algorithmic view or the real view [4,10].

4.2.4.1 The User-Organization's View (Logical View)

The logical view or the user-organization's view for obtaining Se-aaS is the perspective of a sensor-cloud end-user. In Figure 4.3, we depict the logical view of the architecture. The medium of communication between the end-users and the sensor-cloud is a web interface executed at the end of the CSP. Through this interface, user-organizations are authorized to access and request Se-aaS [7]. The access methods are generally controlled by the authorization credentials for every end-user. Once s/he enters the portal, several templates are presented to enable the end-user to specify the requests for sensor data. Additionally, these templates collect information about application types, types of sensor nodes to be used, and the region or zone of interest. The templates enable the end-user to specify requirements of sensor data at a very high level. The templates are then interpreted within the cloud to transform high-level requirements in terms of allocation physical sensor nodes. The subsequent details of this process are discussed later in this chapter.

After specifying the relevant details, the user-organization is ready to enjoy Se-aaS as s/he is completely abstracted from the complicated backend processing logic involved in identification and allocation of the physical sensor nodes, application-specific aggregation, and virtualization. Following the consolidated data processing within the cloud, the user-organization receives the sensed information from the cloud through a portal, and the data are further directed as inputs to the intended application(s).

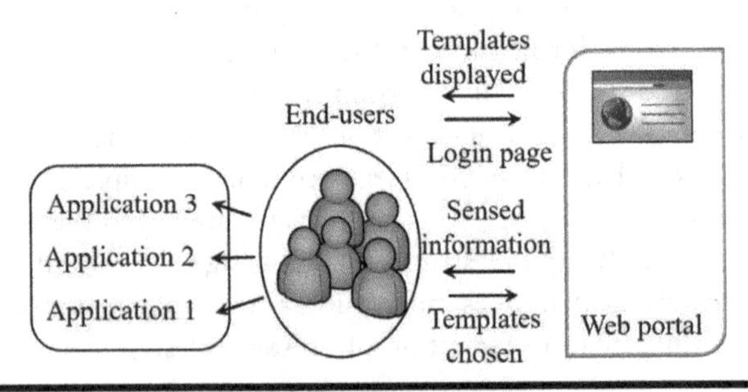

Figure 4.3 User-organization's view of sensor-cloud [10]

4.2.4.2 Algorithmic View (Real View)

Virtualization is the primary component that contributes to the real view of the sensor-cloud, as shown in Figure 4.4. Once the information is obtained and decoded through the templates, the subsequent steps for Se-aaS provisioning are initiated. Every physical node in the sensor-cloud is heterogeneous in terms of the specification, hardware, and configuration. Therefore, the sensor specifications are expressed in a standardized manner using Sensor Modeling Language (SensorML), defined by the Open Geospatial Consortium [7,15]. SensorML is based on XML encoding for ensuring flexible, manageable, and platform-independent processing and analysis of sensor metadata [16]. The process of translating high-level user-requirements, in terms of physical sensor allocation, is one of the major functionalities of the sensor-cloud. Once the physical sensors are allocated to serve a particular application, the data from all of these sensors are virtualized to form the respective VSs. Different VSs serving a particular application are further grouped to form Virtual Sensor Groups (VSGs). After the formation of the VSs and VSGs, the data are pulled from the underlying physical sensor nodes in an on-demand and application-specific manner. The data from the sensors comprising a VS are meaningfully aggregated and dispatched to the respective end-user applications.

Data from every physical sensor node is directed to the cloud infrastructure for storage and processing. Inside the cloud servers, the sensed data are efficiently mined and processed in real time, based on application requirements.

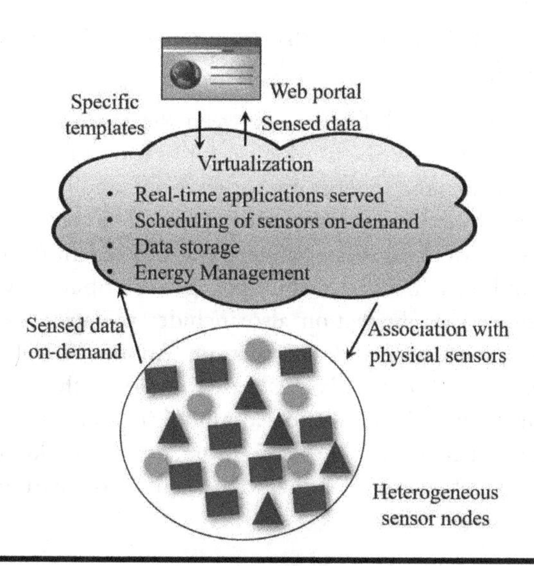

Figure 4.4 Real view of complex processing within the sensor-cloud [10]

The information obtained from a group of physical sensor nodes is further aggregated and sent back to the end-user organization. Thus, an end-user feels that the source of the data is dedicated to his/her application, whereas, practically the set of source sensor nodes is chosen from an infinite pool of resources and changes dynamically to serve multiple applications. Further, the single sensor node may serve multiple applications at a particular time, and the composition of VSs may change based on sensor allocation policies.

Clearly, Figures 4.2 and 4.3 demonstrate the benefits and advantages of the sensor-cloud infrastructure. To sum up, sensor-cloud users can enjoy Se-aaS without worrying about the details of complex processing or other overhead. Therefore, sensor-cloud revolutionizes the perception of sensor network technology to be easy, accessible, seamless, and cost-effective.

Sensor-cloud facilitates the scheduling and execution of customized applications by enabling easier and more effective application switching. This, in turn, ensures not only sharing of sensor services across a bigger scale of users, but also guarantees effective resource provisioning and utilization within the cloud infrastructure as well as within sensor networks. Without the emergence of this paradigm, conventional WSNs would have continued to configure and load applications statically with almost little or no customization. Virtualization is the principal driving force behind this advancement of leveraging the service in a convenient, accessible, beneficial, and adaptable manner for public interests. The pay-per-use policy simply adds to the copious benefits by making the service cheaper and hassle-free.

4.2.4.3 Difference between Sensor-cloud and Virtual Sensor Network

It is important to note the differences between the sensor-cloud and a Virtual Sensor Network (VSN). A VSN is a logical unification of several non-adjacent WSNs that coordinate and cooperate to serve an application with requirements of geo-distributed data processing [17]. The terminology of VSN indicates that the network user is completely abstracted from the complexities of connecting and synchronizing multiple geo-distributed WSNs to serve a single application. The abstraction also includes the spatio-temporal aspects of data transfer, processing, and aggregation. Another significant difference between the sensor-cloud and VSNs, as stated in [4], is that a node in a VSN can be used as an intermediate hop for communication for the sole purpose of connecting multiple networks. On the contrary, in sensor-cloud, virtualization involves the process of grouping nodes that would take part in both sensing and communication.

4.3 Sensor Virtualization

Virtualization of physical sensors makes the paradigm of sensor-cloud unique. As mentioned earlier, virtualization in the sensor-cloud involves the formation and management of multiple VSs [18]. A VS is essentially an assortment of physical sensors from which data are obtained and aggregated. One can visualize a VS serving an application simply as a logical image of the several physical sensors that comprise it.

4.3.1 Configurations of Virtualization

To understand how sensor virtualization works, we first present the different possible configurations of VSs [15,18].

1. *One-to-Many Configuration*: In this configuration, a single physical sensor corresponds to multiple VSs. These VSs may serve multiple applications of different users; however, the data obtained from the particular physical sensors contributes in each of the different VSs that are associated with it. Therefore, the sensor-cloud administrator maintains the sampling frequency and duration of the physical sensors and updates the associated VSs accordingly over time.
2. *Many-to-One Configuration*: As the name indicates, this configuration involves multiple physical sensors contributing to a single VS. This type of configuration is meaningful and applicable if the application to be served demands information such that it has to be acquired from various parts of a geographic region or even across the globe. In such cases, the entire geographic region is divided into sub-regions and sensors from each region together make up a single VS. Whenever data are required from the VS, the corresponding physical sensors are activated and the data are obtained and aggregated. The sampling interval of each of the component sensor must be made identical to that of the query request of the application. This configuration is generally useful for detection of fault tolerance of the underlying WSNs. Several sensors from various locations are sampled at specific times, and a faulty sensor, if any, can be easily identified from the corresponding zone.
3. *Many-to-Many Configuration*: This is essentially a combination of the one-to-many and many-to-one configurations in which a physical sensor can be part of multiple VSs or a component of an assortment for a single VS.
4. *Derived Configuration*: This type of configuration generally abides by the principles of the above three configurations. However, the major difference is that the aforementioned configurations deal with homogeneous sensor types, whereas VSs with derived configurations can contain sensors of

different types. The main utility of this configuration is two-fold. First, this configuration can be used to sense and aggregate data about complex events; second, it can be used to substitute sensors that are not physically present.

4.3.2 Characterization of Virtualization

In this subsection, we focus on the mathematical characterization of the process of virtualization of physical sensor nodes in the sensor-cloud [10]. We start by defining the entities and sub-entities involved in active roles for virtualizing physical sensors.

Every physical node comes with a type mainly describing its sensing hardware. In this model, as presented by Misra *et al.* [10], the type set is described as $T = \{T_1, T_2, \ldots, T_\alpha\}$, where α stands for the number of distinct types of physical sensors. Each sensor $s_i \in S$ is registered to the sensor-cloud by a sensor-owner O_i. If β stands for the number of distinct sensor-owners who have contributed to the sensor-cloud, we express the set of sensor-owners as $O_i \in O = \{O_1, O_2, \ldots, O_\beta\}$. Based on the geographical position of a sensor node and its sensing ability, the location attribute is expressed as $Loc = <l_1, l_2, r>$, where the node is located at (l_1, l_2) coordinate of the globe and has a sensing radius of r units. Therefore, the area of the circle with (l_1, l_2) as the center and r as the radius is the region that is served with the particular sensor. The state of a sensor indicates whether a sensor is currently active (by serving a user-organization) (i.e., $st = 1$) or idle or inactive, (expressed as $st = 0$). The set of cloud service providers is denoted by $CSP = \{CSP_1, CSP_2, \ldots, CSP_\gamma\}$, where γ refers to the overall number of existing cloud service providers in the system. The set of running applications and the set of VS nodes available within the sensor-cloud are denoted by A, and V, respectively.

A physical sensor node is defined as a 7-tuple:

$$s = <id, t, o, Loc, st, csp, QoS>, t \in T, o \in O, csp \in CSP$$

where, $s.id$ is a sensor identification number, locally unique under $s.csp$. The QoS of a sensor node is a composite tuple inclusive of the sensing range of the node, the data transmission range, the energy content, and the accuracy in sensing.

An application $App \in A$ is expressed as a 4-tuple: $App = <A_{id}, A_{type}, A_{sec}, A_{span}>$, where A_{id} is a unique system-generated identification for the application. A_{type} is the type of the application, which will be important for selection of sensors based on types. A_{sec} is a security metric used to specify application requirements in terms of data privacy and protection. A_{span} is the span of the application, expressed as $A_{span} = <Loc_1, Loc_2, Loc_3, Loc_4>$. $Loc_1, Loc_2, Loc_3,$ and Loc_4 are the location coordinates of the four vertices of a rectangular region in sequence, which maps to the span of the application. Based on the application type (A_{type}) and security specifications (A_{sec}), a compatibility function f_1 is built. The goal of the function is to select compatible senor types ($T' \subset T$) and is defined as,

$$f_1(App.A_{type}, App.A_{sec}) = \{T_i : T_i \in T\} = T' \tag{4.1}$$

Once the types of sensor nodes are selected to serve an application, the final set of the physical sensor nodes is selected with a simple allocation function, $f_{alloc}()$. The allocation function is defined as $f_{alloc} : A \rightarrow S_1$. f_{alloc} is a mapping of applications to physical sensor resources S_1, such that $S_1 \in 2^S$. $f_{alloc}()$, in turn, involves a sequence of other intermediate functions $f_1()$, $g_1()$, and $g_2()$, where g_1 is employed to choose a subset of sensor nodes of one or more compatible types. Therefore, we have $g_1 : T \rightarrow 2^S$. $g_1()$ is defined as $g_1(T_j) = \{s_i | s_i \in S, s_i.t = T_j\}$.

The goal of g_2 is to select and assign the location-specific physical sensors (i.e., the chosen sensor nodes must comply and lie within the span of the running application). It is expressed as $g_2 : S_1 \rightarrow S_2$, and $S_1, S_2 \in 2^S$. We have $g_2(S_1) = \{s_k | s_k \in S_1, s_k.Loc \subset App.A_{span}\} = S_2$. Thus, combining the expansions of $g_1()$ and $g_2()$, we obtain:

$$\begin{aligned} f_{alloc}(App) &= g_2(g_1(f_1(App.A_{type}, App.A_{sec}))) \\ &= g_2(g_1(T')) = g_2(\hat{s}, |\hat{s} \in S', S' \subset S, \hat{s}.t \in T') \\ &= \{s \in S_1, S_1 \subseteq S', s.Loc \subset App.A_{span}, \\ s.st &= 0, s.QoS \geq \delta\} \end{aligned} \tag{4.2}$$

where δ is a QoS threshold pre-tuned and pre-decided with the CSP and an end-user-organization. Having understood the allocation function for allocating sensors to applications, mathematically, we will now discuss a mapping $f_{vir} : S \rightarrow V$ expressed as $f_{vir}(f_{alloc}(App_i)) = v_{App_i}$. Every application serving an end-user though sensor-cloud services contains one or more VSs. Thus, $f(App) = v_{App}$. In this case, the application App serves as the input. After computing $f_{alloc}(App) = S_1$, f_{vir} takes S_1 as input. We have,

$$f_{vir}(S_1) = v_{App} | x \in S_1 \wedge x.st = 1 \tag{4.3}$$

$f(App)$ is expressed as,

$$f(App) = y | y \in G, f_{vir}(f_{alloc}(App)) = G = v_{App} \tag{4.4}$$

4.4 Sensor-Cloud Applications

Some of the applications using sensor-cloud platforms are discussed as follows [7]:

1. *Nimbit*: Nimbit is an open-source data processing platform that enables recording and storage of sensor data on the cloud [19]. It is a social service developed for the IoT. Heterogeneous data of various formats (e.g., JSON, GPS, or XML) are obtained within the platform and can be associated with Scalable Vector Graphic processes. Data points can also be controlled to generate patterns, perform analysis, compress, and even generate alerts based on sudden peaks using simple mathematical formulas.
2. *Pachube Platform*: Pachube is one of the first online database service providers [20]. It enables real-time data logging and storage within the cloud platform and provisions a scalable infrastructure for sharing, processing, and collecting sensor data. Pachube comes with various application interfaces to be used within mobile or desktop environments to connect with the cloud servers.
3. *iDigi*: iDigi is a PaaS platform that provides machine-to-machine services for device assets of an organization [21]. For ensuring and catering easy and convenient connectivity within the devices, iDigi presents tools to control, connect, and manage information for an organization. The iDigi Dia software offers to provide information and communication management.
4. *ThingSpeak*: ThingSpeak is an open-source application that enables sensor data collection and storage via LAN or HTTP over the Internet [22]. ThingSpeak offers several basic statistical data processing techniques (e.g., averaging, determining the standard deviation, scaling, rounding off, and so on). ThingSpeak is generally used for target tracking or sensor data logging applications.

4.4.1 Case Studies

Following are case studies for application-specific workflows in the sensor-cloud and WSNs [4,23]. Figure 4.5 narrates a generic workflow representation of any application executed within sensor-cloud platforms [7]. Here, we observe that, when user-organizations need Se-aaS services, they are presented through XML-based templates (i.e., the requests are encoded with XML and are decoded using SensorML interpretations). These templates ask for end-user requirements in a high-level manner. Accordingly, the cloud interprets these requests and deploys the Resource Manager to allocate and reserve physical sensors to serve these requests. The allocation of the physical resources conforms to the definition and the application-specific compatibility of the sensor nodes. To the end-user organization, it appears that it has dedicated sensors to serve its requests anytime, anywhere. This perception is facilitated by the property of virtualization managed by the Virtual Sensor Manager and the Virtual Sensor Controller. They successfully abstract the complexities of the processes involved in resource provisioning, resource management, and data analytics.

4.4.1.1 Target-Tracking Application

A WSN-based target tracking application can be executed for the purpose of intrusion detection (also for surveillance, etc.) Target-tracking applications are generally security-centric. In such a scenario, a WSN-user does not share his/her sensor-based information with an external user, even for business purposes. Therefore, an organization that has its own needs for surveillance or zone monitoring has to deploy its own WSN. As mentioned before, this leads to a heavy payment at the user end for deployment, maintenance, and repairs. Thus, sensor-cloud facilitates provisioning the required Se-aaS to the user-organizations who do not own the WSN [14,11].

Figure 4.6 illustrates how sensors are dynamically allocated based on the position of a moving target. These allocated sensors form VSs to serve the target. Clearly, the compositions of VSs serving such an application varies. The size of a VS is also dependent on the sensing and transmission abilities of the individual sensors constituting the VS. The VS is kept alive as long the application keeps executing, after which it is killed, and the sensors are deallocated. The

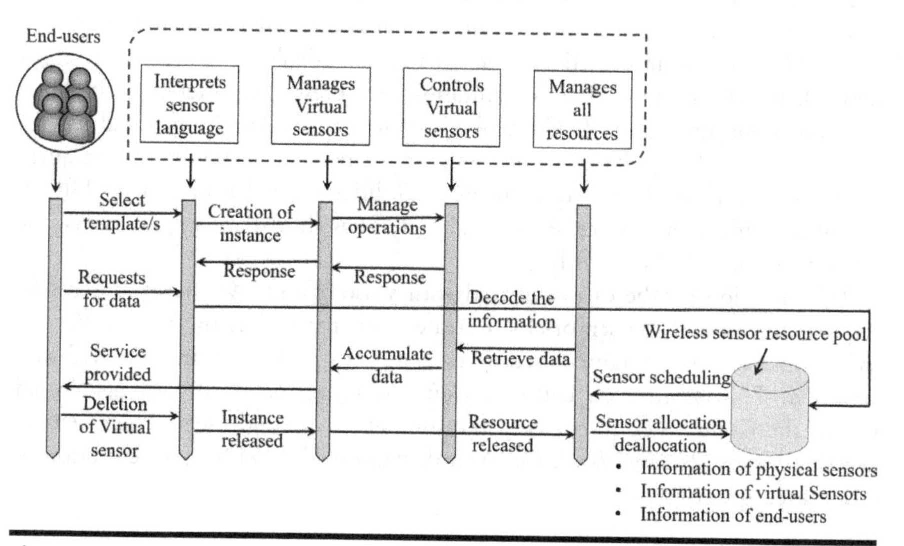

Figure 4.5 Workflow in Sensor-cloud [10]

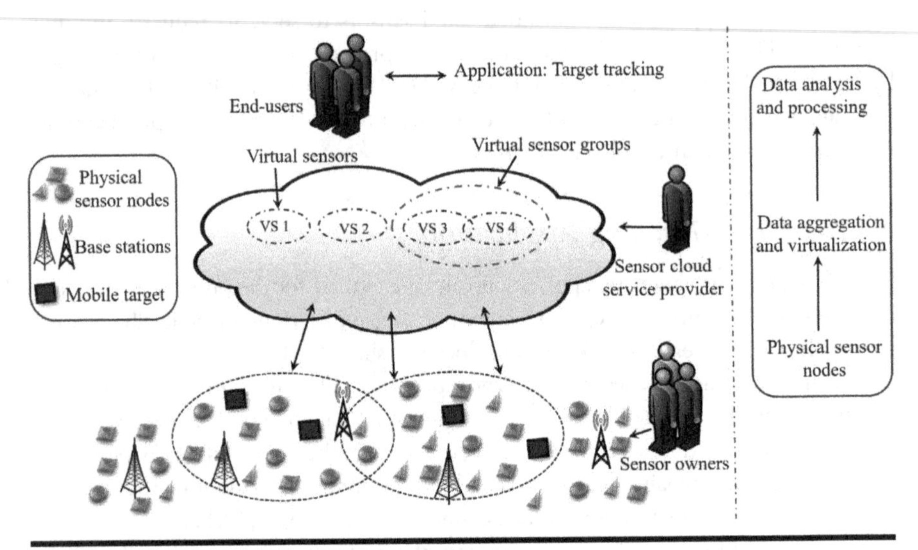

Figure 4.6 Target tracking in sensor-cloud [11]

Virtual Sensor Manager is responsible for the real-time and efficient data processing and analytics thereby enabling the user-organization to enjoy Se-aaS.

4.4.1.2 Weather Services

We present a use case of sensor-based environment monitoring for weather services. Let us assume that user-organization *A* is interested in rainfall, temperature, and humidity information. The organization deploys its very own sensor network for the purpose. User-organization *B* expresses its interest in building business connections with *A*. The goal of *B* is to retrieve data from *A* and pay for the service accordingly. Now *A* will configure the network based on its own query requirements. Thus, if the requirements of *B* differ from that of *A*, it is difficult to put up with such services. Additionally, if *A* has short-term requirements, the investment on a WSN is costly.

This situation can be easily resolved with sensor-cloud. Whenever *A* issues its requests for Se-aaS, the sensor-cloud creates and instantiates the correct VS. All the necessary sensors (rainfall sensor or temperature sensor) are allocated and activated. The information collected from the group of sensors are aggregated within the cloud and subsequently transmitted to the end-users. These sensors can simultaneously serve *B* with its own demand of data with its specification of customized querying.

4.5 Some Insights

As we observe, sensor-cloud infrastructure is considered a substitute for conventional WSNs, it is imperative to experimentally justify the necessity for this shift of paradigm for all WSN-based applications to a sensor-cloud platform. Therefore, in this section, a detailed analysis is presented based on the performance metrics of the sensor-cloud; a comparative study is performed with conventional WSNs. From a network point of view, the energy consumption of the nodes and the fault tolerance of the network are studied and analyzed. From a business point of view, the cost-effectiveness of the sensor-cloud is examined by determining the characteristics of cash inflow and outflow of every actor of the sensor-cloud. The experimental setup of this work is illustrated in Table 4.1.

4.5.1 Performance Metrics

Misra *et al.* [10] defined the following metrics to evaluate the performance of sensor-cloud systems compared to traditional WSNs:

(i) *Energy Consumption*: The consumption of energy E is the summation of the energy expenses due to transmission (E_{tr}), reception (E_r), sensing (E_s), and computation (E_{proc}). The units of energy consumption are kept

Table 4.1 Experimental setup [10]

Parameters	*Values*
Time period	5 simulation years (60 simulation months)
Deployment area	500 m × 500 m
Deployment	Uniform, random
Number of sensor nodes (N)	1,000
Number of sensor-owners (η_1)	5
Number of end-users (η_2)	10
Unit cost price of a node (C_s)	20 currency unit
Unit cost due to deployment (C_{deploy})	3 currency unit/sensor
Unit cost due to maintenance ($C_{maintain}$)	10 currency unit/month
Unit cost due to rent (C_{rent})	10 currency unit/month
Cost per unit usage of Se-aaS (C_{Se-aaS})	10 currency unit/month
Communication range	[50, 100] m
Transmission energy (E_{tr})	7 nJ/bit
Computation energy (E_{proc})	5 nJ/sec
Sensing energy (E_s)	6 nJ/event
Time interval for nodes being faulty (Ω)	5

identical for each energy component for both WSN and the sensor-cloud. Therefore,

$$E = E_{tr} + E_r + E_s + E_{proc} \quad (4.5)$$

(ii) *Fault Tolerance*: As per [10], we have the following definition: "Fault-tolerance, $\mathcal{F}$ of a network is defined as the total number of non-faulty nodes present in the network at a particular time. Mathematically,

$$F_t = F_{t-1} - P_f \times F_{t-1}, \mathcal{F}_0 = N \quad (4.6)$$

where N and P_f are the total number of operative nodes initially present in the network and the percentage of faulty nodes, respectively."

(iii) *Cost-efficiency*: For the analysis of cost-efficiency of the sensor-cloud, we present a study on the cash flow of every actor and compare it with that of a WSN user. Plots represent the cumulative cash flow of every actor over time. Plots along the negative y-axis represent a cash outflow CO (or a loss) from the actor, whereas plots along the positive y-axis represent cash inflow CI (or gain) to the actor. The respective costs due to deployment, maintenance, and rent are denoted by C_{deploy}, $C_{maintain}$, and C_{rent}, respectively.

a) Sensor-owner: For a sensor-owner, the cash flow is as follows:

$$CO_{sensor-owner} = n_1 \times (C_s + C_{deploy}) \quad (4.7)$$

$$CI_{sensor-owner} = n_1 \times C_{rent} \quad (4.8)$$

where, n_1 is the total number of sensors registered by the sensor-owner at C_s unit cost price of each sensor node.

b) End-user:

- WSN user: For a WSN user,

$$CO_{wsn} = n_2 \times (C_s + C_{deploy} + C_{maintain}) + n_3 \times C_{deploy} \quad (4.9)$$

where n_2 represents the total count of nodes in the WSN and n_3 refers to the number of faulty nodes. A WSN user is basically served in terms of the sensed data, and there is no cash inflow for such user.

c) Sensor-cloud end-user: For a sensor-cloud end-user who enjoys Se-aaS, the cash outflow is as follows:

$$CO_{end-user} = n_4 \times C_{Se-aaS} \tag{4.10}$$

Here, n_4 represents the number of sensor nodes that have been utilized by the user in obtaining Se-aaS on a monthly basis. C_{Se-aaS} is the corresponding cost to the user for each unit of Se-aaS consumed monthly.

d) CSP[2]: For a CSP, the monthly inflow and outflow of cash are shown below:

$$CO_{csp} = \eta_1 \times CI_{sensor-owner} + \ \Omega \times n_5(C_{deploy} + C_{maintain}) \tag{4.11}$$

$$CI_{csp} = \eta_2 \times CO_{end-user} \tag{4.12}$$

where η_1 is the number of sensor-owners registered with the system, η_2 stands for the number of end-users served, and Ω is the time period or the frequency at which periodic maintenance activities have to be performed at the cloud end. After Ω duration of time, n_5 nodes turn out to be faulty or need replacement and maintenance.

4.5.2 Performance Evaluation

The analysis is performed separately for each of the metrics indicated in Section 4.5.1.

We will first start with the performance analysis of sensor nodes in terms of their energy consumption. In Figure 4.7(a), the cumulative energy consumption of WSNs and sensor-cloud over time are studied and plotted. The total energy consumption is broken down into the consumption due to computation and data transmission. As mentioned before, in sensor-cloud, a significant amount of computation is offloaded to the powerful cloud servers; hence, the energy consumed due to computation significantly decreases. Also, as WSNs are based on in-network computation, it leads to a large number of packet transmissions within the network for computation through broadcasting, negotiation, cluster computation, decision making, and so on. Therefore, the energy consumption due to data transmission is less in sensor-cloud than in WSNs. This, in turn, affects the total energy consumption trend as well by reducing the consumption by 36.68% in sensor-cloud.

Next, we present the performance analysis for the fault-tolerance metric in sensor-cloud. In Figure 4.7(b), we observe the variation in the number of faulty nodes for both the sensor-cloud and WSNs. In the era of the IoT, fault-tolerance is extremely crucial in sensor networks. For WSNs, if nodes turn out to be faulty at a given rate, the network eventually reaches a dead state, given that no redeployment or replacement takes place to substitute for the faulty nodes. In

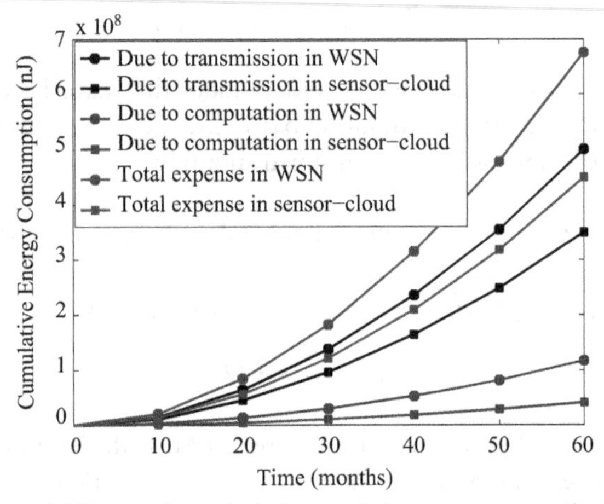

(a) Comparative analysis for cumulative energy consumption

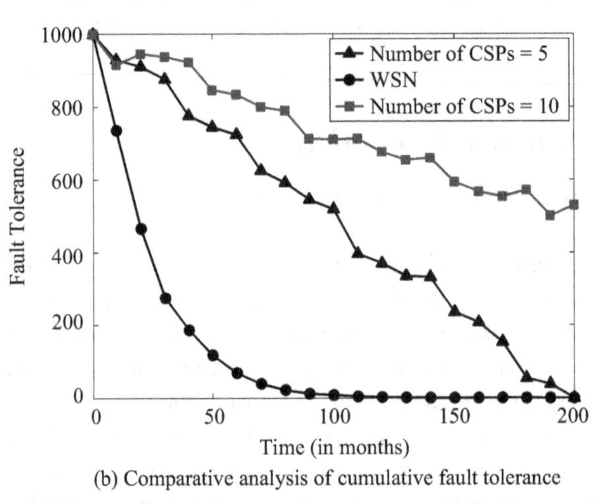

(b) Comparative analysis of cumulative fault tolerance

Figure 4.7 Comparative performance analysis of sensor-cloud and WSN [10]

the sensor-cloud, this problem is largely mitigated as end-users have the opportunity to choose the "best" sensors from any service provider. Therefore, it is the responsibility of the service provider to select the appropriate sensors to best serve the user. Again, in Figure 4.7(b), there is an analysis on the performance improvement with the increase in the number of CSPs. With more CSPs, the likelihood of fault-tolerance is less as the load of application requests is distributed across multiple CSPs, thereby preventing the over-use or exhaustion of the same node.

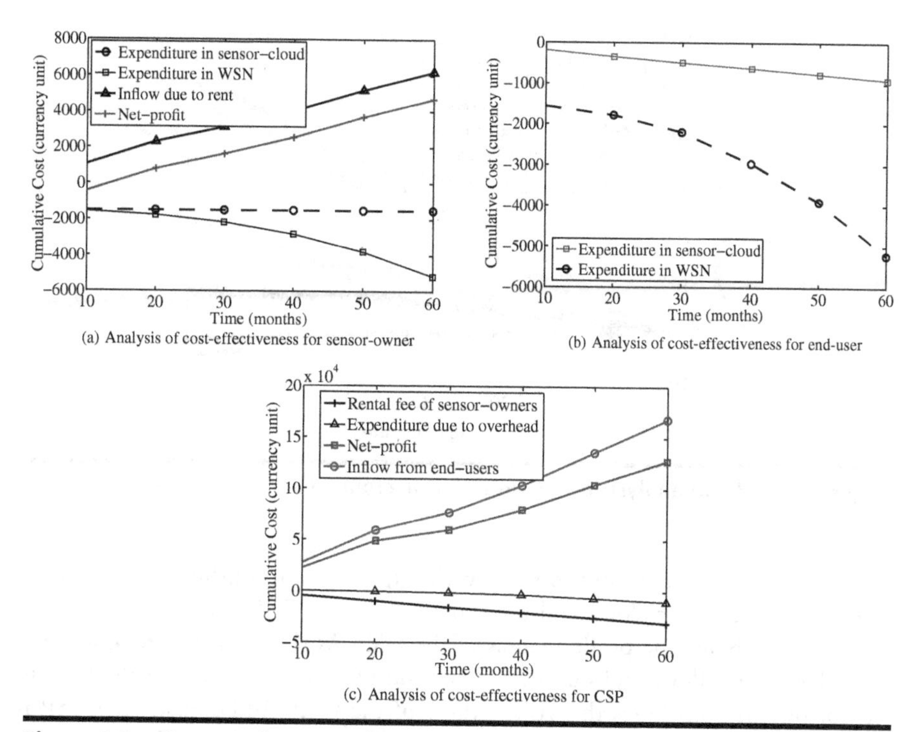

(a) Analysis of cost-effectiveness for sensor-owner

(b) Analysis of cost-effectiveness for end-user

(c) Analysis of cost-effectiveness for CSP

Figure 4.8 Comparative cost-effectiveness analysis of sensor-cloud and WSN [10]

Having discussed the performance evaluation of the network parameters, we now present the cash flow analysis of the different actors in the sensor-cloud. Figure 4.8(a) highlights the cash inflow and outflow of sensor-owners. In a WSN, the WSN user is the sensor-owner and the data consumer, purchasing, deploying, maintaining, and managing the network personally. As we can observe, the cash outflow happens only once for sensor-owners, for the purchase of sensors, which is a one-time investment. However, there is a continuous cash inflow for the sensor-owner based on the monthly rental profit obtained by the CSP. Therefore, analytically, we conclude that the sensor-owner enjoys 33.83% more profit in sensor-cloud compared to a WSN user.

For the end-user, the cash flow analysis is very different. In Figure 4.8(b), the flow of cash is highlighted for the end-user. A WSN-user is responsible for every activity related to the sensor network. A sensor-cloud end-user obtains the service, does not worry about the other aspects, and is charged only for the units of Se-aaS consumed. Therefore, there is no such cash inflow for the end-user; the profit is manifested in terms of the service. Again, as the user is free from

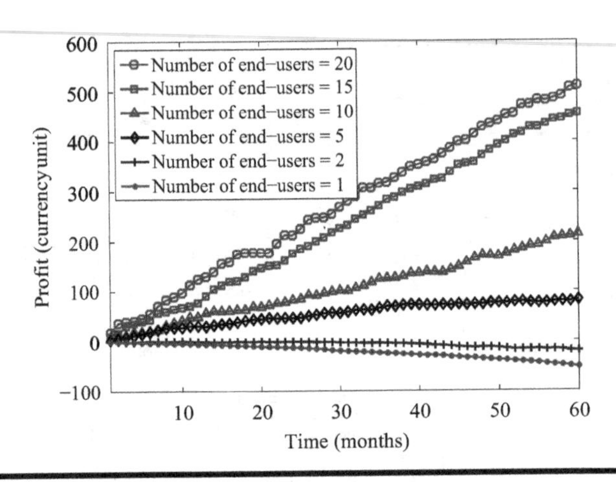

Figure 4.9 Profit analysis of CSP in a sensor-cloud [10]

maintenance activities and other overhead, the cash outflow decreases by 14.72% for a sensor-cloud end-user compared to a WSN user.

We now discuss the profit analysis of the CSP. As shown in Figure 4.8(c), the CSP has cash inflows and outflows. The monthly rental-fee paid by the CSP to sensor-owners constitutes the cash outflow of the CSP. Additionally, the CSP is responsible for periodic maintenance and redeployment of the physical sensor nodes. In turn, the cash inflow of the CSP is the profit enjoyed in terms of the cash outflow of the end-users for buying Se-aaS. The figure indicates the effective profit of the CSP and how it cumulates over time.

The sensor-cloud infrastructure cannot be efficiently used for any kind of sensor services. It is advisable to use such platforms when the requested resource types are available and usable. For example, if an end-user requests Se-aaS from a not-so-common (or less-demanded) sensor type, it will incur a high overhead and maintenance at the cloud end, which is likely to surpass the profit of the CSP for selling Se-aaS, as shown in Figure 4.9. We observe that for serving Se-aaS to one or two end-users with resources in low demand, the CSP incurs loss. Only when the demand increases (i.e., the number of users requesting such resources increase) does the CSP incur profit.

4.6 Summary

In this chapter, we discussed the concepts and principles of the sensor-cloud paradigm, that is considered a substitute for traditional WSNs. Starting from the background of sensor-cloud and the reasons for its inception, we discussed the motivation behind such a paradigm. We illustrated the details of the different actors of the sensor-cloud, the architecture that it follows, and the logical and the actual view of the sensor-cloud. As mentioned before, sensor virtualization is the backbone of the sensor-cloud; hence, we thoroughly illustrated the principles of virtualization inclusive of the various potential configurations of virtualization and a theoretical characterization of virtualization. Then we discussed a few application-specific examples of the sensor-cloud and showed a case study to understand the workflow. Lastly, to justify the paradigm shift, we discussed some experimental results and an analysis for the performance of sensor-cloud platforms when compared to WSNs.

Working Exercises

Numerical Problems

1. If a cloud server can support up to 10 VSs, what kind of configuration will you use for (a) 5 sensors, (b) 10 sensors, and (c) 20 sensors?
2. An application area spans a rectangle positioned around (0,0), (0,100), (100, 100), and (100, 0). What would be the communication radius of the sensor serving the application?
3. A network comprises 1,000 physical sensors with a daily fault rate of 0.23. When will the network die?
4. A WSN has 1,000 nodes, out of which 20 are faulty. The unit cost of a node is Rs. 10. The unit cost due to deployment is Rs. 8, and the unit cost due to maintenance is Rs. 5. What is the cash outflow of a WSN user?
5. What is the cash outflow for a sensor-owner for the same setup if he owns 10 sensors?

Conceptual Questions

1. Who are the actors of the sensor-cloud and what are their roles?
2. What is the key tier in sensor-cloud architecture and why?
3. What are the different virtualization schemes in the sensor-cloud?
4. What is the sequence of various operations within sensor-cloud for a target tracking application?
5. Why is the sensor-cloud energy efficient compared to conventional WSNs?

Notes

1 www.sensorcloud.com/system-overview

2 The third actor in sensor-cloud infrastructure is the sensor-cloud administrator, essentially a non-human actor and hence not a participant in the economics of the model. However, although the CSP is not a potential actor in the infrastructure, it is the most significant business entity of the sensor-cloud; hence, the cash flow analysis of the CSP is of high interest.

References

[1] C. Zhu, H. Wang, and X. Liu, "A Novel Sensory Data Processing Framework to Integrate Sensor Networks with Mobile Cloud," *IEEE Systems Journal*, vol. 10, no. 3, pp. 1125–1136, Aug. 2016.

[2] K. M. C. Kumar, "Internet of Fitness Things A Move towards Quantified Health: Concept, Sensor-Cloud Network, Protocols and A New Methodology for OSA Patients," *IEEE Recent Advances in Intelligent Computational Systems (RAICS)*, 2015, doi: 10.1109/RAICS.2015.7488443.

[3] A. Alamri, W. S. Ansari, M. M. Hassan, M. S. Hossain, A. Alelaiwi, and M. A. Hossain, "A Survey on Sensor-Cloud: Architecture, Applications, and Approaches," *International Journal of Distributed Sensor Networks*, vol. 2013, Nov. 2013, doi: 10.1155/2013/917923.

[4] K.-L. Tan, "Whats NExT?: Sensor + Cloud!?" in *Proceedings of the 7th International Workshop on Data Management for Sensor Networks*, ACM, New York, NY, 2010, pp. 1–1.

[5] L. Ferretti, F. Pierazzi, M. Colajanni, and M. Marchetti, "Scalable Architecture for Multi-User Encrypted SQL Operations on Cloud Database Services," *IEEE Transactions on Cloud Computing*, vol. 2, pp. 448–458, Oct. 2014.

[6] X. Zhang, L. Yang, C. Liu, and J. Chen, "A Scalable Two-Phase Top-Down Specialization Approach for Data Anonymization Using MapReduce on Cloud," *IEEE Transactions on Parallel and Distributed Systems*, vol. 25, no. 2, pp. 363–373, Feb. 2014.

[7] K.-W. Park, J. Han, J. Chung, and K. H. Park, "THEMIS: A Mutually Verifiable Billing System for the Cloud Computing Environment," *IEEE Transactions on Services Computing*, vol. 6, no. 3, pp. 300–313, 2013.

[8] C. Yang, C. Liu, and X. Zhang, "A Time Efficient Approach for Detecting Errors in Big Sensor Data on Cloud," *IEEE Transactions on Parallel and Distributed Systems*, Jan. 2014, doi: 10.1109/TPDS.2013.2295810.

[9] S. Chatterjee, R. Ladia, and S. Misra, "Dynamic Optimal Pricing for Heterogeneous Service Oriented Architecture of Sensor-Cloud Infrastructure," *IEEE Transactions on Services Computing*, Jul. 2015, doi: 10.1109/TSC.2015.2453958.

[10] S. Misra, S. Chatterjee, and M. S. Obaidat, "On Theoretical Modeling of Sensor Cloud: A Paradigm Shift from Wireless Sensor Network," *IEEE Systems Journal*, 2014, doi: 10.1109/JSYST.2014.2362617.

[11] S. Chatterjee and S. Misra, "Target Tracking Using Sensor-Cloud: Sensor-Target Mapping in Presence of Overlapping Coverage," *IEEE Communications Letters*, vol. 18, no. 8, pp. 1435-1438, Aug. 2014.

[12] S. Chatterjee, S. Misra, and S. U. Khan, "Optimal Data Center Scheduling for Quality of Service Management in Sensor-Cloud," *IEEE Transactions on Cloud Computing*, 2015, doi: 10.1109/TCC.2015.2487973.

[13] S. Chatterjee and S. Misra, "Optimal Composition of a Virtual Sensor for Efficient Virtualization within Sensor-Cloud," in *Proceedings of IEEE International Conference on Communications (ICC)*, 2015, doi: 10.1109/ICC.2015.7248362.

[14] M. Yuriyama and T. Kushida, "Sensor-Cloud Infrastructure – Physical Sensor Management with Virtualized Sensors on Cloud Computing," in *Proceedings of the*

13th International Conference on Network-Based Information Systems (NBiS), Sept. 2010, pp. 1–8.

[15] Open Geospatial Consortium. www.opengeospatial.org/.

[16] M. Botts, Ed., Sensor Model Language (Sensorml) for In-Situ and Remote Sensors, Open Geospatial Consortium Inc., 2004. Online: https://www.opengeospatial.org/standards/sensorml.

[17] A. P. Jayasumana, Q. Han, and T. H. Illangasekare, "Virtual Sensor Networks – a Resource Efficient Approach for Concurrent Applications," in *Proceedings of the 13th International Conference on Information Technology*, 2007, pp. 111–115.

[18] S. Madria, V. Kumar, and R. Dalvi, "Sensor Cloud: A Cloud of Virtual Sensors," *IEEE Software*, vol. 31, no. 2, pp. 70–77, Mar. 2014.

[19] Nimbits Data Logging Cloud Sever, www.nimbits.com.

[20] Pachube Feed Cloud Service, www.pachube.com.

[21] iDigiDevice Cloud, www.idigi.com.

[22] IoT ThingSpeak, www.thingspeak.com.

[23] S. Misra, A. Singh, and S. Chatterjee, "Mils-Cloud: A Sensor-Cloud-Based Architecture for the Integration of Military Tri-Services Operations and Decision Making," *IEEE Systems Journal*, 2014, doi: 10.1109/JSYST.2014.2316013.

[24] S. Misra, S. Bera, and T. Ojha, "D2P: Distributed Dynamic Pricing Policy in Smart Grid for PHEVs Management," *IEEE Transactions on Parallel and Distributed Systems*, vol. 25, 2014, doi: 10.1109/TPDS.2014.2315195.

Chapter 5

Data Flow in the Sensor-Cloud

5.1 Introduction

This chapter presents the details of data flow and management in the sensor-cloud. As we know, data are transmitted to the end-user from a virtual sensor (VS), and each VS is served using multiple sensor nodes and cloud data centers. Hence, it is crucial to understand the formation of VSs specific to an application.

As the underlying sensor nodes are highly resource-constrained, the entire network and cloud performance may be severely affected by unoptimized and inefficient utilization of resources. In this chapter, we illustrate the process of optimal composition of a VS for efficient virtualization within the sensor-cloud. Next, we discuss the procedures for overall data management and flow from WSNs to the sensor-cloud. We discuss how the data transmission scheme from the underlying WSNs occurs and how the data are managed within the sensor-cloud. Thus, this chapter covers specifically two distinct aspects related to sensor-cloud: composition of a VS and data management.

5.2 Composition of a Virtual Sensor

We know that physical sensors are allocated based on the application demand. These sensors are further grouped logically to form a VS. These VSs are grouped to form Virtual Sensor Groups (VSGs). Whenever an end-user requests for Se-aaS, it is provisioned through the data acquired from VSs or VSGs [1,14]. It is important to know the process for composing a VS. The basic and naive approach is to select the sensors for an application and allocate all of them to constitute the VS. This approach is called the maximal composition approach. However, as the nodes are resource-constrained, we cannot always afford to follow the maximal composition approach, especially when the number of selected nodes is very high. Therefore, the composition of every VS should be optimal. For this, we focus on algorithms that ensure the dynamic and optimal formation of VSs for different applications.

The rationale behind the optimal composition of a VS is backed up by the fact that the lifetime of the battery-driven sensor nodes is limited, after which the nodes must be charged or replaced. In Chapter 4, we learned that sensors are allocated to applications based on compatibility, defined in terms of the type, location, and Quality of Service (QoS) of the application and other application-specific requirements. From this set of sensors, we have to choose the optimal subset. This subset has to be the minimal subset, which is able to meet the QoS demands of the application. Any larger subset results in redundant resource utilization and consumption.

Our goal, therefore, is to choose the optimal set of sensors, that is not only compatible to serve an application, but also has the ability to maintain the efficiency of resource utilization. Based on the works of Chatterjee and Misra [1,2], we discuss the following algorithms: (i) Composition of VS within the same region (*CoV-I*) and (ii) Composition of VS and VSG across multiple regions (*CoV-II*). The CoV-I algorithm helps compose a VS with homogeneous sensor nodes, which are chosen from a single geographic region (as shown in Figure 5.1 (a)); while the CoV-II algorithm focuses on the composition of a VS with heterogeneous geo-distributed sensors that eventually make up a VSG (as shown in Figure 5.1(b)).

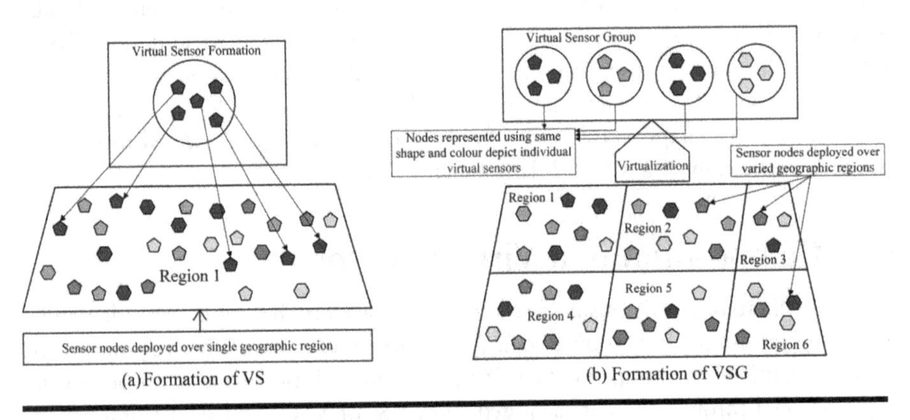

Figure 5.1 Virtualization of the physical sensor nodes [1]

5.2.1 Algorithms for Optimal Composition

The following discussion details the differences between the two aforementioned cases and how the two algorithms cater to the different needs.

Case (a): We begin with Figure 5.1(a), which highlights the composition of a VS for case (a). We observe that the physical sensors selected to serve the application are co-located within the same geographical region. These sensors are homogeneous in nature in terms of their sensing and communication hardware. For this case, *CoV-I* is proposed, which is subsequently discussed in Section 5.2.1.

Case (b): In this case, as shown in Figure 5.1(a), the sensor nodes selected for the application belong to several geographical regions (R_1, R_2, R_3, R_4, R_5, and R_6). Thus, different geo-distributed VSs are selected to serve the application. These VSs comprise nodes that are heterogeneous in nature in terms of their sensing and communication hardware. For this case, the *CoV-II* algorithm, which is discussed in Section 5.2.2, is used.

Before we get into the formation of the mathematical optimization, we present some key terms to be used in our discussion as stated in [1]. We consider that physical sensors nodes are deployed across r non-overlapping finite regions, $\mathcal{R} = \{\mathcal{R}_1, \mathcal{R}_2, \ldots, \mathcal{R}_r\}, \mathcal{R}_i \bigcap \mathcal{R}_j = \Phi, \forall 1 \leq i,j \leq r, i \neq j$. The information about a sensor known *a priori* is the location of its deployment (l_1, l_2), representing the latitude and longitude of the absolute position of the node and its sensing radius at time t, λ_t. The sensors nominated to serve an application *App* spanning over $\mathcal{R}_{req}$ ($\mathcal{R}_{req} \subseteq \mathcal{R}$) as the region of interest is given by,

$$\mathcal{S} = \{s_i\} \mid (\varphi(l_1, l_2, \lambda_t^{s_i}) \subseteq \mathcal{R}_{req}) \wedge (s_i.type = App.type). \qquad (5.1)$$

φ is a method that accepts the sensor information as input and returns the region that falls under the sensing range of the nodes. The "*type*" attribute of sensors corresponds to what it does (e.g., rainfall sensor and temperature sensor). If two sensor nodes s_i and s_j are homogeneous, $s_i.type = s_j.type$ holds true.

5.2.2 CoV-I: Composition of VS within the Same Region

We now discuss the *CoV-I* algorithm for the composition of a VS comprising homogeneous nodes belonging to the same geographical region (i.e., $\mathcal{S} = \{s_i\}$), where $s_i.type = s_j.type$ and Equation (5.1) holds true. Let us assume that the sensor-cloud has received K application requests and $\mathcal{P}_i$ is the priority of App_i[1]. Let the set obtained from maximal composition be $\mathcal{S}$. It is important to measure the "goodness" $\mathcal{G}$ of every physical node s_i in $\mathcal{S}$. We present the definitions of the factors affecting the "goodness" of a node, as defined by Chatterjee and Misra [1]:

Definition 1. The normalized residual energy of a sensor node is defined as the ratio of the current energy level to the initial energy level, expressed as, $\mathcal{Q}^t_{s_i} = \frac{E^{cur}_{s_i}}{E^{init}_{s_i}}$, where $E^{cur}$and E^{init} the current and the initial battery levels of the node, respectively.

Definition 2. The expected received signal strength intensity [17] at time t is defined as the expected value of the signal strength when perceived at the cloud end at t.

Definition 3. Proximity of s_i with a base station (BS) is given by, $\chi_{s_i} = \sqrt{(s_i.x - BS.x)^2 + (s_i.y - BS.y)^2}^{-1}$, where x and y represent the abscissa and the ordinate, respectively.

The goodness metric $\mathcal{G}$ can be formulated using the methodology discussed in the work of Chatterjee and Misra [1]. The authors also modeled the Quality of Information (QoI) of a sensor node s_i. The use of the model is inspired from the work of Ciftcioglu *et al.* [3]. The authors maximized the QoI using Lagrangian multipliers on the constraints and optimized the composition of the VSs for varying execution priorities of end-user applications [4].

The authors maximized the QoI using Lagrangian multipliers on the constraints and optimized the composition of the VSs for varying execution priorities of end-user applications.

5.2.3 CoV-II: Composition of VS and VSG across Multiple Regions

We present the *CoV-II* algorithm that considers several types $T = \{t_1, t_2, ..., t_f\}$ of compatible sensor nodes $\mathcal{S} = \{s_i\}$ distributed over several regions. If the composition of a VS includes sensors of type $t_f \in T$, $\forall s_j \mid s_j.type = t_f$, the primal of the algorithm is formed with Lagrangian multipliers [1]. Thus, the VSGs $\mathcal{V}_i$ for App_i, $i = 1, 2, ..., K$ using t_j sensor types, are expressed as,

$$\mathcal{V}_i = \{\Omega^*_{App_i,t_1}, \Omega^*_{App_i,t_2}, \dots, \Omega^*_{App_i,t_f}\} \tag{5.2}$$

These VSGs include those VSs that contain the membership of physical sensor nodes selected optimally in terms of their resource ability, as mapped to the priority of applications. The analyses of this work indicates that the losslessness of the proposed algorithms, *CoV-I* and *CoV-II.*

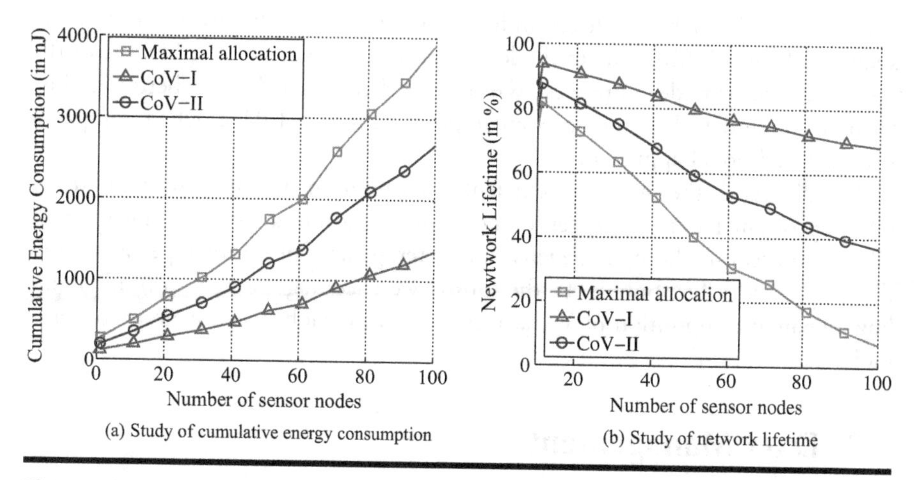

Figure 5.2 Comparative study of the network parameters [1]

5.2.4 Performance Evaluation of CoVs

We show the performance evaluation of *CoV-I* and *CoV-II* algorithms. We present the experimental setup [1] in Table 5.1.

Figure 5.2 illustrates a comparative analysis of the network performance of *CoV-I* and *CoV-II* with the approach of maximal composition. The parameters considered are the cumulative energy consumption and network lifetime. For analysis of cumulative energy consumption, we have included the energy consumption components due to transmission and reception of data, sensing of events, and computation and analysis. The analysis from Figure 5.2(a) reveals that the energy consumption using the maximal composition is 34.9% higher compared to *CoV-I*, whereas it is 68.4% more than *CoV-II*. Clearly, the use of the lower number of sensor nodes is one of the main reasons behind low energy

Table 5.1 Experimental Setup [1,2]

Parameters	*Values*
Deployment area	500 m × 500 m
Deployment type	Uniform, random
Number of nodes	100
Communication energy	70 nJ/bit
Energy due to computation	30 nJ
Sensing energy	10 nJ/bit
Number of applications	3
Application priority	{1, 2, ..., 10}

consumption. This has a direct influence on the network lifetime as well, as shown in Figure 5.2(b). Due to the lower energy consumption, the lifetime of the individual sensor nodes improves. We observe that *CoV-I* shows a better lifetime by 61.04%, and *CoV-II* indicates an improvement by 61.04%, when compared with the traditional approach.

We discussed the issue of efficient virtualization within sensor-cloud by selecting optimal number of sensor nodes to constitute a VS. Existing research takes into account the homogeneity and heterogeneity of sensor types as well as the geographical distribution. In the future, we encourage readers to think about how dynamic composition of VSs can be done with mobile sensor nodes to serve mobile applications.

5.3 Data Management

The three-tier architecture of sensor-cloud [5] was presented in Figure 4.2. In this architecture, we observe that the physical sensor nodes are at the bottom-most layer followed by the cloud infrastructure in the middle layer and the end-user organizations at the top. The user requests are captured, decoded, and served by the cloud infrastructure with the use of sensor services from the underlying network.

5.3.1 Data Caching

In the sensor-cloud architecture, we observe that, based on the demand of an application, end-users directly access the physical sensor network through the sensor-cloud. To put it in another way, whenever an application needs to be served with Se-aaS, the corresponding physical sensors are identified and activated and data are retrieved from the sensors subsequently. If an application has a high demand, data are pulled more frequently from the sensor networks than those with low demand. Consequently, the components of a VS transmit more or less based on the application requirement. However, as an application varies, so the nature of the sensed environment also varies. If we take the example of a rainfall-measuring sensor, the sensed information does not change frequently at consecutive instants, whereas for a camera sensor mounted on a busy street, the sensed data frames change several times every second. In the former case, a data-demanding application will lead to continuous data pulling and data transmission from the rainfall sensor, even if the variation of the sensed data is insignificant due to a low rate of change in the environment. This not only incurs a huge overhead for repeating the process of redirecting every query to the lowest tier and fetching the data up to the uppermost tier; it also increases the energy consumption and reduces the network lifetime, in turn. Herein, we observe the significance and utilization of data caching in the sensor-cloud.

A major aspect of data management in the sensor-cloud is data caching, which we will cover in this section. We illustrate the dynamic data caching mechanism in the sensor-cloud that improves the energy consumption and network lifetime and preserves the accuracy of information as well. The proposed data-caching mechanism is sensitive and adaptive to changes in the environment. Based on the changes, the idea is to decide either to pull sensed data from the networks or to serve the end-users with cached information without really affecting the information accuracy. In this regard, Chatterjee and Misra [6] proposed a dynamic and optimal data caching policy that determines an optimal data-caching interval between two consecutive data pulls from the actual sensor network. The proposed solutions claim to reduce the energy consumption of the network and improve the battery life consequently.

5.3.1.1 Architecture for Data Caching

The main motivation of data caching in the sensor-cloud is that the continuous access to underlying sensor networks could lead to redundant sensing and data transmission, especially for sensors whose environment does not change frequently. This is detrimental from the perspective of resource utilization in sensor networks, as shown in Figure 5.3(a). In this section, we explain the proposed cache-enabled architecture of the sensor-cloud.

Figure 5.3(b) highlights the existence of two different caches [6]: the *primary cache* or the *external cache* (*EC*), and the *secondary cache* or the *internal cache* (*IC*). The EC is positioned between the sensor network and the cloud infrastructure,

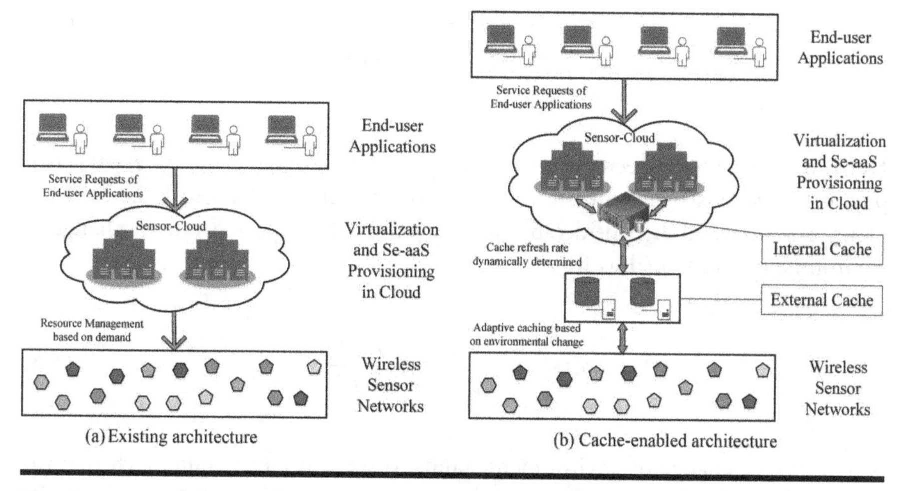

Figure 5.3 Existing and proposed architectures of the sensor-cloud [6]

thereby capturing the dynamism or the change of the physical environment. Based on the changes of the sensed environment, EC modifies its content and stays updated with the physical environment. On the other hand, the IC stands between the EC and the end-users. It is ideally located with the respective VMs of the users and caches the data based on the application the VM is serving. Each IC on a VM updates its content in synchronization with the content of the EC.

5.3.1.2 The Model of the External Cache

As the EC is located between the physical sensor networks and the sensor-cloud, the objective that supports the functionality of the EC is to determine the largest caching interval Δt for any sensor s [6]. For this purpose, it is assumed that we have the history of the previous k readings $R_1 = \{r_1, r_2, \ldots, r_k\}$ at the corresponding times of $T_1 = \{t_1, t_2, \ldots, t_k\}$. With this data at hand, the goal is to determine the next caching instant k', so that $k' - k$ can be as large as possible, subject to the constraints to be discussed later.

Definition 4. The current memory m of EC, at time t, is a k tuple, where k is a pre-negotiated system value. m is expressed as [6],

$$m(t) = \{(r_1, t_1), (r_2, t_2), \ldots, (r_k, t_k)\}, r_i \in R_1, t_j \in T_1 \tag{5.3}$$

Therefore, the mean rate of change of environment, e, is expressed as,

$$e = \frac{\sum_{i=2}^{k} | m(t).r_i - m(t).r_{i-1} |}{\sum_{i=2}^{k} m(t).t_i - m(t).t_{i-1}} \tag{5.4}$$

Definition 5. The expected rate of change of environment, e, for a particular physical sensor, is based on the th^i degree of the rate of change of the environment, $1 \leq i \leq k$. After considerable simplification, the expected rate of change of environment is obtained as [6]

$$E(e) = c_1 \frac{dR_1}{dT_1} + c_2 \frac{d^2 R_1}{dT_1^2}. \tag{5.5}$$

The rate of change of environment stands to serve as a constraint such that the magnitude of $E(e)$ has an upper bound $e_{threshold}$, beyond which there is a necessity to re-cache data and update the content. For the energy expenditure, specific costs are associated with different means of energy consumption – transmission of data packets E_{tr}, reception of instruction or data packets E_r, and per state transition from passive (idle or not transmitting) to active (transmitting) and vice versa E_{st} [7]. With these energy components, the final energy consumption is modeled [6]. As transmission energy involves a round trip to send and receive information, a cost factor of 2 is associated with E_{tr}. Based on this, the energy minimization problem is expressed as,

$$\text{Minimize} \sum_{t_i=2}^{k'} (2\alpha E_{tr} + \beta E_{st} + (t_i - t_{i-1})\gamma E_s).$$

The above optimization problem is solved to obtainthe optimal value of k'^*.

Therefore, the optimal caching interval $\Delta k^* = k'^* - k$ is determined in such a way that the overall energy consumption of the network is minimized and the cache is updated with the rate of change of the environment. At the end of the caching interval, the process of cache updating is initiated. The interval may also dynamically change over time, based on the environment and its rate of change.

5.3.1.3 The Model of the Internal Cache

Coming to the second level of cache, which is the IC, we observe that the IC resides within the VMs of every end-user. Thus, it is primarily responsible for handling application requests from end-users [6]. For every request, the IC determines whether the data for the request should be retrieved from the network or from the cache. Data retrieved from the sensor network is simultaneously cached within the IC. Here, we assume that $d = \{d_i\}, 1 \leq i \leq p$ is the data stream provisioned at time $\{t_i\}$. Thus, the p data packets constitute the memory of the IC. The expected rate of change of EC, e', is expressed as

$$E(e') = \frac{\sum_{j=2}^{k} \mid d_j - d_{j-1} \mid}{\sum_{i=2}^{k} t_i - t_{i-1}}. \tag{5.6}$$

Definition 6. The mean accuracy $\hat{A}$ of data provisioning is defined as the inverse of the Mean Square Error (MSE) of the sensor readings at the EC and IC, evaluated for the previous j time instants. Thus, the mean accuracy of a data at time t is expressed as [6],

$$\hat{A} = \frac{j}{\sum_{i=t-j+1}^{t} (m(t).r_i - d_i)^2} \quad (5.7)$$

In order to ensure accuracy, the following needs to hold true:

$$\sum_{i=t-j+1}^{t} (m(t).r_i - d_i)^2 \rightarrow r, r \rightarrow 0, r \neq 0. \quad (5.8)$$

If k be the last time instant at which data has been cached within the IC and k'' is the next instant for caching data, the overall minimization problem for IC is as follows:

$$Maximize(k'' - k),\ i.e., Minimize\ g_1(k') = \frac{1}{k'' - k}, \quad (5.9)$$

subject to the constraint

$$\sum_{i=k''-j+1}^{t} (m(k'').r_i - d_i)^2 - r \approx 0. \quad (5.10)$$

Therefore, solving the equations, k'' is determined as

$$k \in \min_{k<h<g} \left\{ \max \left\{ \sum_{i=h-j+1}^{h} (m(h).r_i - d_i)^2 \right\} \right\}. \quad (5.11)$$

The value of k'' specifies the time at which the IC will be updated with the particular sensor value of the application. After k'', this process is reiterated for the next values to be cached within the IC.

5.3.1.4 The Performance of Caching

We now discuss the performance of the aforementioned data-caching policy in the sensor-cloud. Table 5.2 summarizes the experimental setup [6]. The first experiment focuses on the study and justification of the proposed methods of computation of $E(e)$ and $E(e')$ using Equations (5.5) and (5.6), respectively. The experiment involves capturing randomized sensor data for 100 time instants. For this purpose, applications are classified into two distinct categories: (i) stable (i.e.,

Table 5.2 Experimental setup [6]

Parameters	*Values*
Deployment area	500 m × 500 m
Deployment	Uniform, random
Number of nodes	100
Transmission energy (E_{tr})	7 nJ/bit
Energy due to state transition (E_{st})	30 nJ
Sensing energy (E_s)	6 nJ/event
Number of time instants	100

applications in which the data gradually change but the standard deviation between two consecutive data readings is never too high) (see Figure 5.4(a)) and (ii) unstable (i.e., applications in which the data fluctuate very frequently from very low to very high values) (see Figure 5.4(b)). In Figure 5.4(a), we observe a smoothness in data variation, which helps in an accurate estimation of the change of environment. On the other hand, Figure 5.4(b) reflects turbulence in data variation; thus, the estimation has a greater deviation from the original data as compared to the former case.

The definition of data accuracy is given in Definition 6. It is computed with the help of RMSE. In Figure 5.5(a), we observe that the error is larger when the rate of change of environment is comparatively higher in Figure 5.4(a). However, with the stability in the environment, the error for estimation of the change of environment is almost negligible for both the EC and IC. The error observed in the first couple of time instants are comparatively higher due to the continuous learning within the EC and the IC, before the estimation could be done with precision. On the contrary, Figure 5.5(b) exhibits a higher error because of the turbulence in the environment. This implies that the proposed learning fits well only for stable applications.

The accuracy of computation, as given in Definition 6, is evaluated in terms of the computation of the RMSE for the above two scenarios [6]. Figure 5.5(a) clearly shows the RMSE obtained for expecting the rate of change of an environment in a stable condition. The error obtained for the first few time instants is initially high, due to the gradual learning or adaptiveness of the caching process, after which the error falls to a negligible value. For an unstable environment, as depicted in Figure 5.5(b), the RMSE in expecting the change of the environment rises and falls sporadically, based on the rate of change of the environment.

For studying the energy conservation through the proposed caching mechanism, the cumulative energy consumption, $\mathcal{E}$, is given as [6],

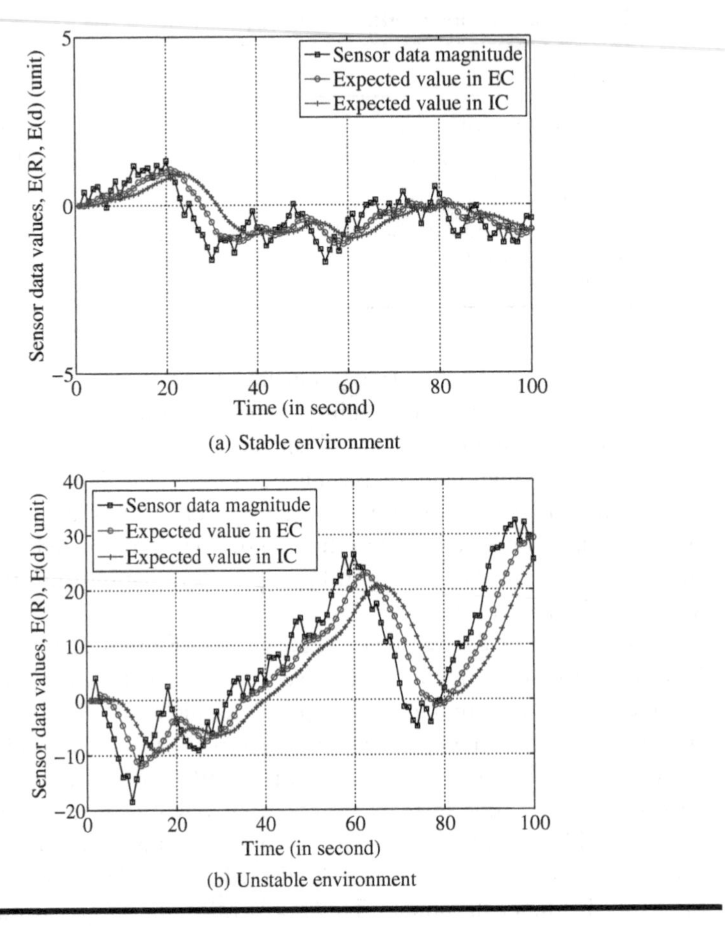

Figure 5.4 Study of the expectation of the sensed data with time [6]

$$\forall 1 \leq j \leq 100, \mathcal{E}(j) = \sum_{t_i=1}^{j} (2\alpha E_{tr} + \beta E_{st} + (t_i - t_{i-1})\gamma E_s). \tag{5.12}$$

Figure 5.6(a) compares the energy consumption of the proposed caching mechanism with the conventional non-caching ones. The conventional benchmark used for experimentation focuses on frequent data transmission without data caching [4]. Clearly, the energy consumption due to data packet transmission is higher than that of the proposed method. This also reduces the overall energy consumption by 37.1%, thereby supporting the requirement for data caching. Figure 5.6(b) is also obtained as a direct consequence, which reflects the improvement in the overall network lifetime $\mathcal{N}$. At any given time, t, we have,

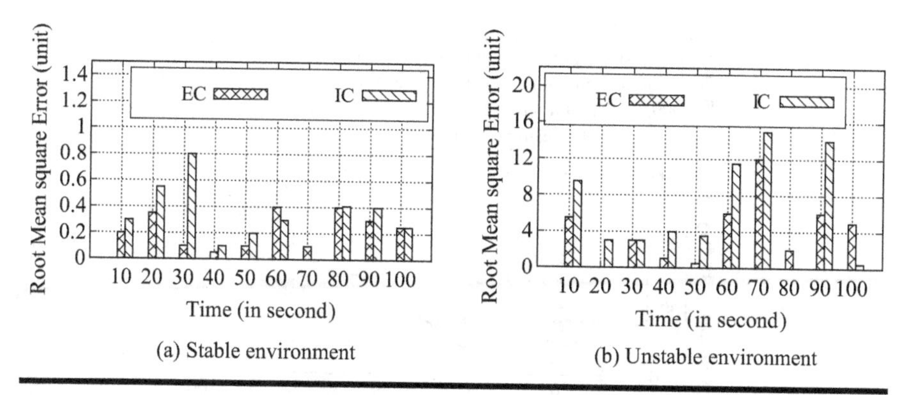

Figure 5.5 Analysis of the RMSE in computation of the expectation of sensed data [6]

$$\mathcal{N}(t) = \frac{\mathcal{N}_{max} - \mathcal{E}(t)}{\mathcal{N}_{max}} \times 100\% \tag{5.13}$$

with the maximum life of a node $\mathcal{N}_{max}$ in terms of energy taken as $6000nJ$. We observe an improvement of approximately 48% by using environment-aware data caching policies.

From another perspective, we present Figure 5.7, which shows the evaluation of the caching mechanisms within the EC and IC. In the stem plots, a value of 0 represents the time instant where cached data has been used, whereas a value of 1 is indicated when the cache has been updated with modified content. In Figure 5.7(b), we observe the fluctuation of the original sensor data and how the EC has been refreshed to update the content with the change in the environment. In Figure 5.7(a), however, we observe how changes in the EC trigger the cache updating in IC. The study is performed for 100 time instants to evaluate the refresh rate of the data in both the EC and IC.

5.3.2 Data Transmission

We now focus on the data transmission aspect of the sensor-cloud, which involves efficient routing and forwarding of data from the sensor networks to the cloud. Motivated by the works of Chatterjee *et al.* [2], the objective of this section is to familiarize the readers with the following:

(i) The hierarchy of physical sensor nodes serving as intermediate hops during data transmission.

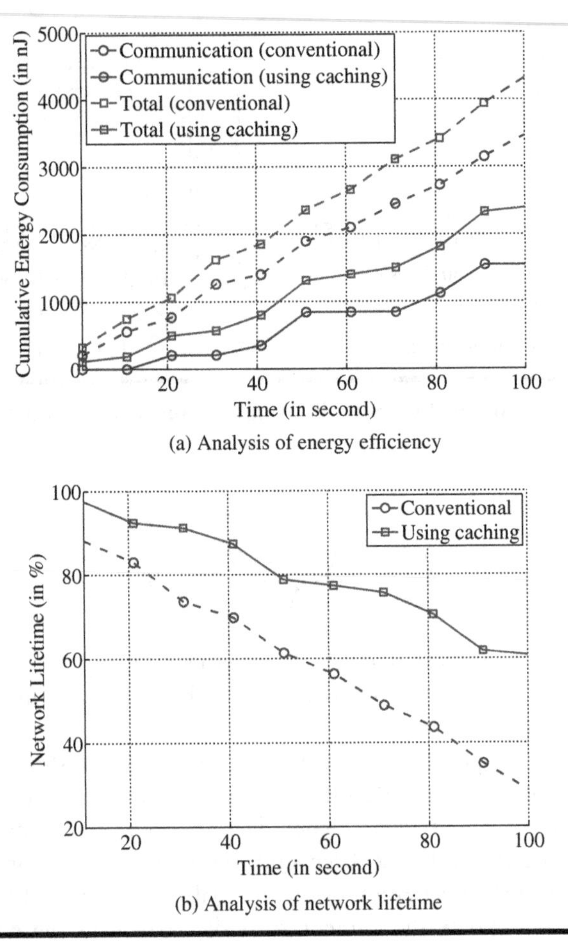

Figure 5.6 Overall analysis of the network resources [6]

(ii) Ensuring efficiency in the process of data transmission, in terms of energy and cost-effectiveness.

By this time, we are aware of the heterogeneity of sensor nodes in the underlying network of the sensor-cloud. Unlike WSNs, in which the property of heterogeneity is mainly observed for sensing, communication, and variation of battery life [8–10], in the sensor-cloud, the nodes exhibit heterogeneity in terms of hardware and specifications, in addition to the aforementioned parameters. This makes the data transmission aspects extremely challenging and difficult from sensor networks to cloud. Another point of difference

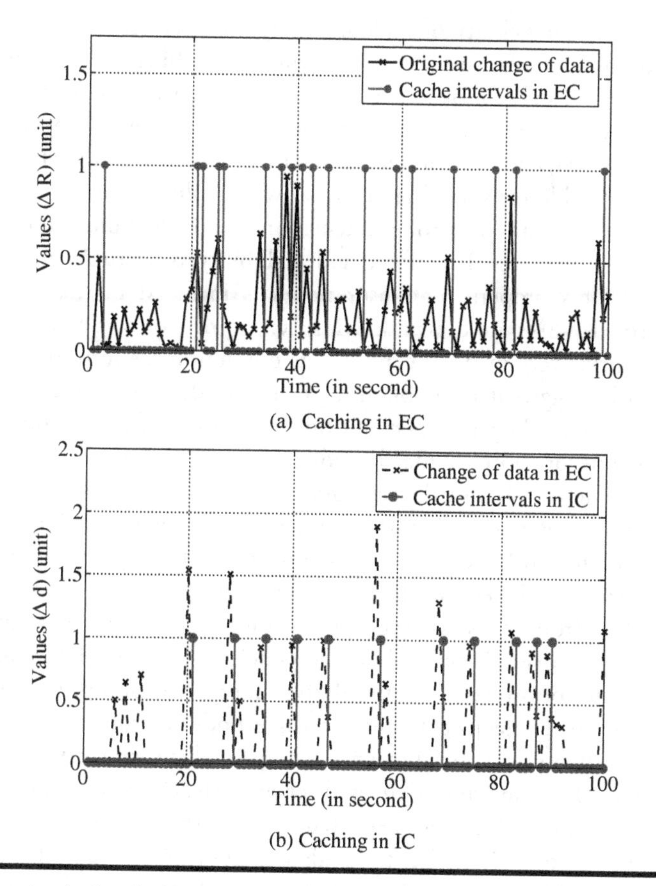

Figure 5.7 Analysis of adaptiveness and dynamism of caching [6]

between sensor-cloud and WSNs is the proximity of sensor nodes. In conventional WSNs, a sensor node of a particular type is deployed within the direct reach of another node of the same type, which is not true in the case of the sensor-cloud. Here, nodes serving a particular application may not be within the communication range of the other nodes serving the same application, thereby giving rise to virtual sensor networks (VSNs) [11] within the physical networks.

This work focuses on introducing a hierarchy of sensor nodes for energy-efficient data transmission from sensor networks to the cloud. It might seem that this problem has already been addressed in routing sensor data from the networks to the base station, in the case of the traditional WSNs. However, to date such solutions do not take into account the probability of a node turning out to be erroneous (i.e., the possibility of node fallibility). Therefore, one may ask, "How can a sensor node misbehave as it is essentially nothing more than a simple computational device?" Sensor nodes are indeed designed to respect the internal algorithms they are fed. However, multiple factors influenced by the external environment can change the normal behavior of the node. In the era of the IoT, a node can be fallible for several reasons: software malfunctions; a part of certain code that fails, leading to wrong calculations or deductions in the decision making of the node, etc. Therefore, node fallibility is a crucial metric to take into account. This property of sensor nodes is included in this work. In the logically formed VSNs in the underlying sensor networks, decisions are often made in network; every member expresses its opinion in the form of data packets, which are further aggregated to obtain the collective decision of the network. The aggregated decision is hugely dependent on the decision of every individual node. Therefore, if one or more of the decisions from the sensor nodes vary, it may change the overall decision of the network [2].

This section presents the multi-hop data transmission scheme from the physical sensor networks to the sensor-cloud [2]: (i) This section explains the problem of energy-efficient multi-hop data routing within VSNs and transmission of data from VSNs to the sensor-cloud. The proposed solution takes into account the overall network heterogeneity. (ii) The data transmission scheme aims to optimize the cumulative energy consumption of a VSN, while selecting or deselecting an intermediate node for routing. (iii) The data transmission scheme is associated with an optimal decision rule that chooses the particular decision rule that will ensure the best selection of sensor nodes, while maximizing the expected payoff from a VSN in terms of energy consumption. (iv) The transmission schemes proposed in [2] take into account fallibility in the decision making of nodes, which is extremely relatable in the current era of the IoT, when a node has a likelihood of being vulnerable to privacy threats and security breaches. The decision-making ability of a node is expressed as the probability to select or deselect another node correctly.

5.3.2.1 *Architecture for Data Transmission*

In Figure 6.1, we observe the multiple tiers of routing involved in the sensor-cloud. The lowest layer specifies the VSN. A set of sensor nodes is chosen from the VSN to serve an application. Within this set, only a small number of nodes make it to the list of what we call the "first hop bridge nodes." The first hop bridge nodes contain information collected from the VSN. Depending on the energy content and the proximity of the bridge nodes to the base station, the next subset is formed; this group is called the final hop bridge nodes. These nodes are responsible for transmitting the data directly to the base station, as these nodes are directly within the reach of the base station. Figure 6.1 demonstrates a three-tier architecture with two layers of bridge nodes. However, in practical situations, the layers of bridge nodes can be more than two, depending on the number of physical nodes within the VSN.

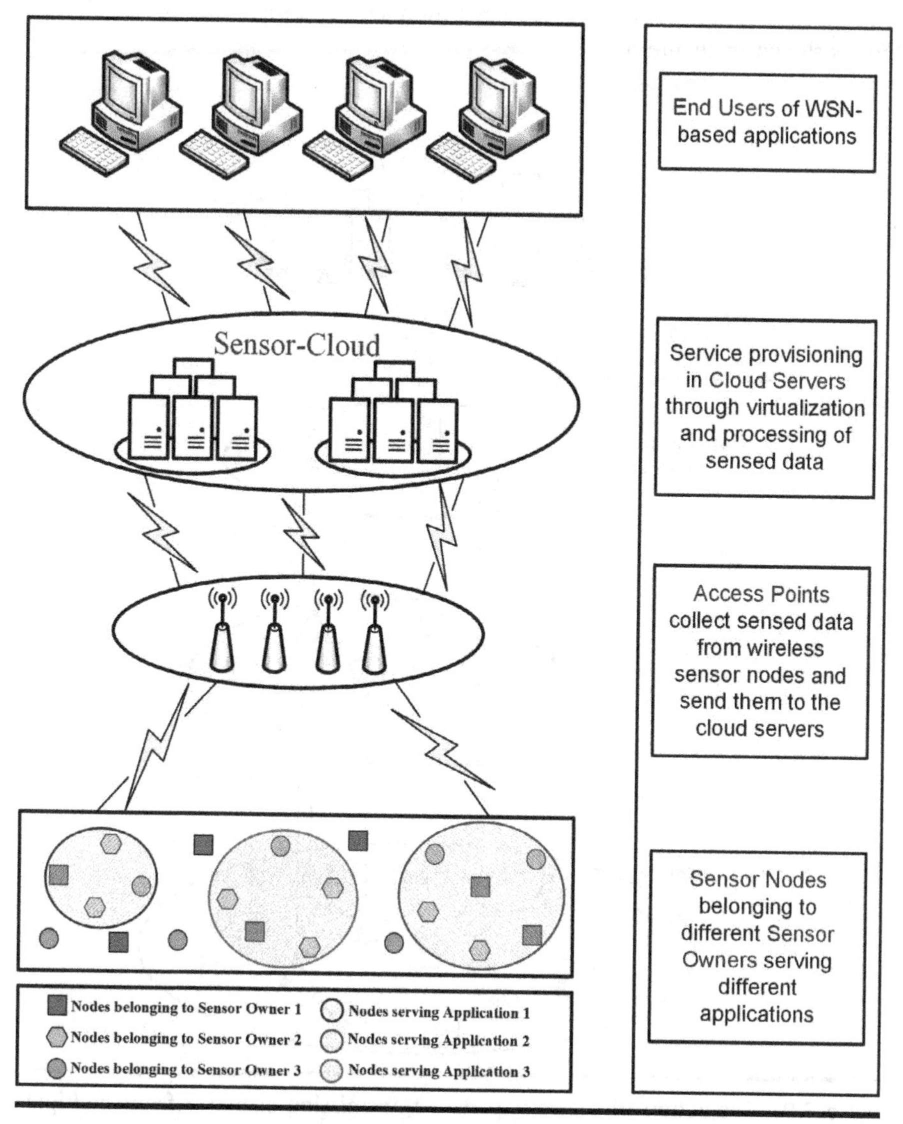

Figure 5.8 Network architecture of Sensor-cloud [2]

Several VSNs are considered within the physical sensor networks formed and maintained to serve multiple applications. The basic routing scheme within the network is considered to be multi-hop to the cloud-end [12,13,15]. However, the selection of nodes varies based on several parameters: power consumption, node lifetime, and network lifetime. Also, with a vision to improve the overall network performance, the logical hierarchy of nodes is introduced for multiple levels of bridge nodes. The nomenclature is influenced by the fact that such nodes function as a bridge connecting the large number of sensor nodes to the cloud-end, as shown in Figure 5.9.

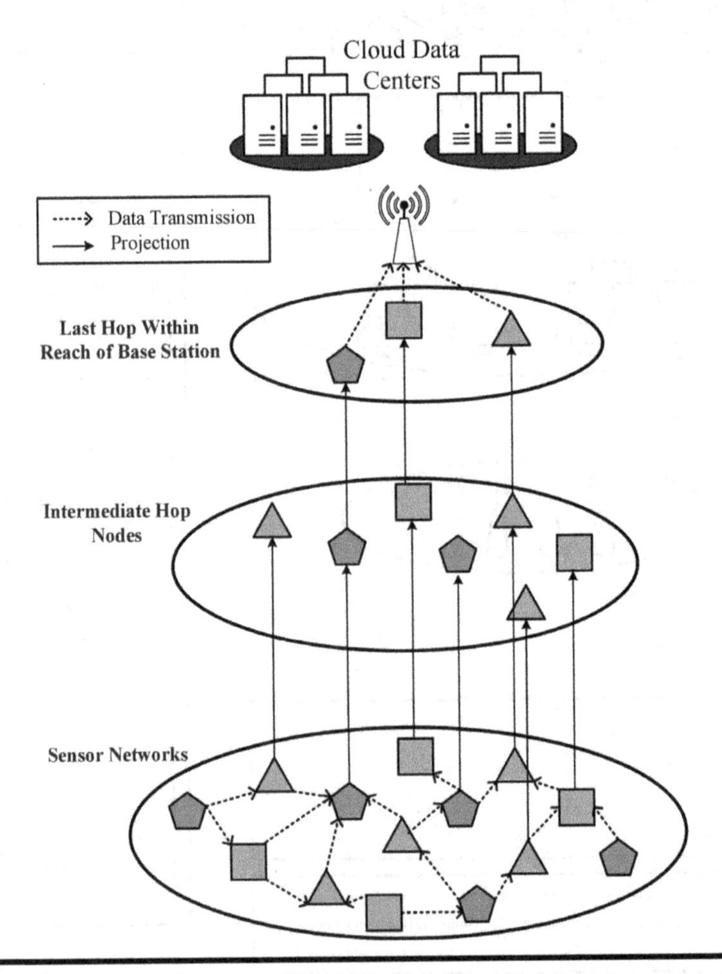

Figure 5.9 Projection of a two-hop data transmission scenario from multiple VSNs [2]

5.3.2.2 Formal Definition of the Transmission Scheme

We now present the formal definition of the problem to be discussed. We consider the total number of physical nodes within the WSN to be n, which, in turn, constitutes the VSN for a particular end-user and an application.

We consider that n number of physical sensor nodes are registered within the sensor-cloud infrastructure [2]. A subset of these nodes is grouped together to form a VSN comprising virtual sensors that are used to serve a particular end-user application. The components of this system are formally defined in Table 5.3.

Now the goal is to efficiently choose the particular node that will serve as the bridge for a VSN for a single iteration. Selection of such a node will vary with every iteration, as using a node for too long will exhaust its energy content and the goal is to choose the node with the best configuration. The set of potential bridge nodes is indicated as $BN = \{b_1, b_2, \ldots, b_m\}, m \ll n, \; \exists i,$ such that b_i is the winner bridge node. The objective function is a mapping from a VSN to a single bridge node. Therefore, mathematically, we have $f : V \rightarrow BN$. This is an optimization problem for minimizing the power consumption of the VSN. As stated in [2], f belongs to the solution set of the minimization function,

$$\arg\min \sum_{\forall s_i \in \nu_k} E_{ij} \tag{5.14}$$

$$\text{where, } E_{ij} = m_i r_j E_{i,elect} + m_i \xi_{ij}^2 \epsilon_{i,amp} \tag{5.15}$$

Table 5.3 Notation of Parameters used [2]

Parameters	*Explanation*
$S = \{s_1, s_2, \cdots, s_n\}$	Set of physical nodes available
$V = \{\nu_{k,u_j}\},\ 1 \leq k \leq 2^n$	Set of VSNs
ν_{k,u_j}	k^{th} VSN of end-user u_j
ξ_{ij}	Euclidean distance between s_i and s_j
E_{elect}	Energy dissipated due to transmitter or receiver circuitry
E_{amp}	Energy dissipated due to transmission of amplifier
m_i	Message transmission rate of s_i
r_i	Duration of the ith round
E_{ij}	Energy loss of s_i for the jth round

The optimization problem has been solved in the existing literature using the method of *Optimal Decision Rule* [15] in a "general pairwise choice framework" setting. The node fallibility is expressed as a probability with which a bridge node can be selected or deselected depending on its metric of "goodness" or "badness".

5.3.2.3 Performance of the Transmission Scheme

We now present the results of the data transmission policy within the sensor-cloud. For evaluation of performance, a comparison of the proposed technique with the random bridge node selection algorithm as well as the centralized bridge node system is presented to investigate the performance. The experimental setup of the parameters is elaborated in Table 5.4 followed by the discussion of the performance evaluation [2].

We begin with an analysis of the cumulative energy consumption of a WSN and how it changes over time. Figure 5.10 highlights the energy consumption using the three different schemes: the optimal decision rule, random selection, and the centralized bridge node selection method. For the random node selection method, any node could serve as the bridge node. This selection does not reflect selection of nodes based on proximity to the base station or the energy content of a node. Therefore, it may lead to the selection of a node that is far from the base station, thereby incurring significant energy expense. In the centralized bridge selection method, all nodes of the VSN transmit data packets to a single sink node. The approach is multi-hop, but centralized, thereby ensuring significant energy loss, as shown in Figure 5.10. The optimal decision rule carefully chooses the bridge nodes based on the metrics of energy and distance. This selection is also dynamically updated based on the change in the magnitude of the metrics. Therefore, the energy consumption is the least in this method. In Figure 5.11, we demonstrate the energy variation as expended by a non-bridge node. The

Table 5.4 Experimental Setup [2]

Parameters	*Values*
Deployment Area	100 m × 100 m
Deployment	Uniform, random
Number of nodes	100
Channel overhead	[1, 5]%
E_{elect}	50 nJ/bit [8]
E_{amp}	100 pJ/bit-m^2 [8]
Packet size	2358 bytes

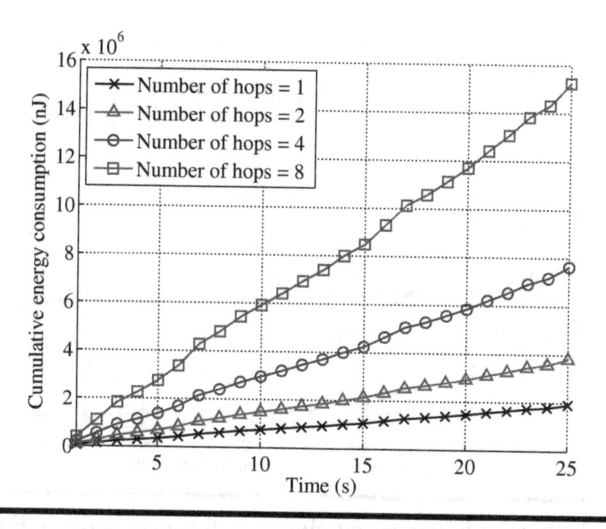

Figure 5.10 Cumulative energy consumption of a VSN vs. time [2]

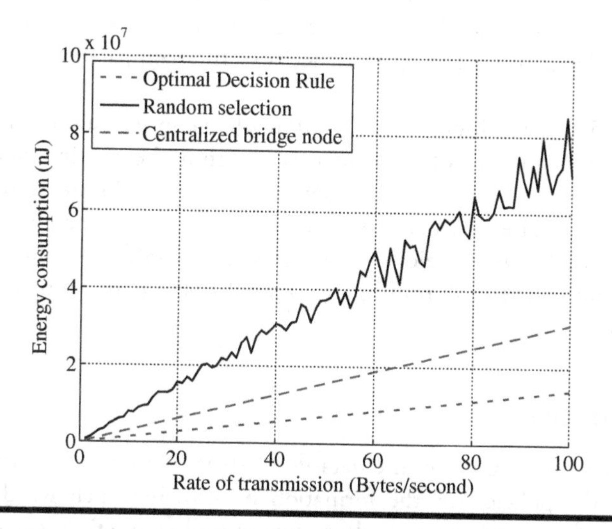

Figure 5.11 Energy consumption for a node vs. rate of data transmission [2]

energy consumption naturally increases with the transmission rate of data packets.

In Figure 5.12, we observe the network lifetime of non-bridge nodes over time. The conservation of network lifetime depends on the energy-efficient selection of the bridge node. As the number of nodes increases, the lifetime decreases. This

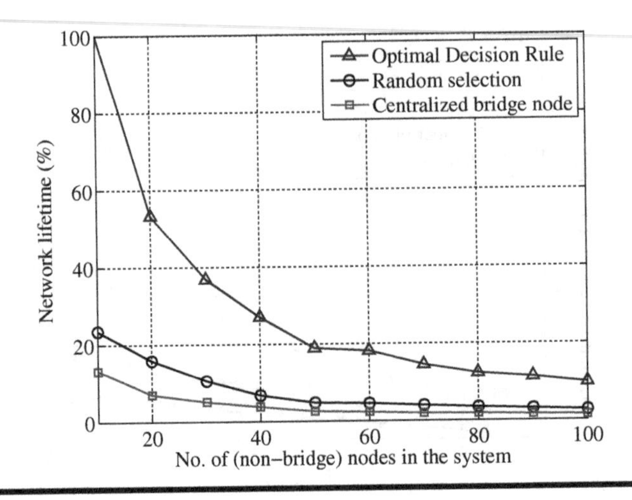

Figure 5.12 Network lifetime vs. number of non-bridge nodes in the system [2]

indicates the significance of introducing the hierarchy of bridge nodes for conservation of resources. We also observe that the lifetime of the network is significantly higher using the optimal decision rule compared to the other two methods.

In Figure 5.13, we discuss the variation of energy consumption of a VSN over time as the number of hops increases within the routing scheme. With the increase in the hop count, the energy expenditure due to transmission of data packets increases, which, in turn, increases the energy transmission overhead as well. Nevertheless, it is always advisable to choose between centralized and distributed routing schemes, based on the topological design of the network.

5.4 Summary

In this chapter, we focused on the data-flow aspects within the sensor-cloud. We started with the policies for the formation of VSs, in which we discussed the composition of a VS within a single region as well as multiple non-overlapping regions. Next, we discussed the data management issues of the sensor-cloud. We covered the dynamic and optimal data caching schemes that ensure accuracy of information. We also presented details about energy-efficient data transmission policies from sensor networks to the sensor-cloud for efficient and reliable communication.

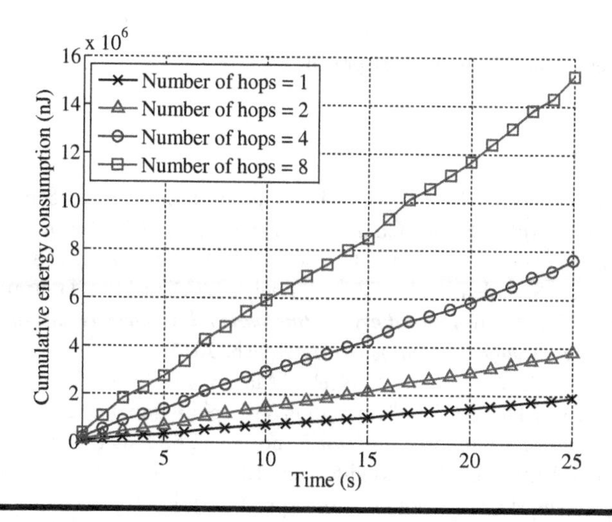

Figure 5.13 Cumulative energy consumption vs. hop count of bridge nodes [2]

Working Exercises

Numerical Problems

1. The dynamism of a physical environment is captured as follows:

$$D = 3x^4 + 7x^3 + 32x + 6$$

What is the expected rate of change of the environment, if the constant to capture the change is 0.3 at x=1?

2. Sensor data from a physical environment reported the following values from t_1 to t_{15}:$\{2, 2.3, 3, 2.4, 5, 6, 2.1, 2.9, 3.2, 7, 2, 10, 11, 2.7, 2.9\}$. When should the cache be refreshed if the mean is set to 5?
3. At a particular time, when the sensor reads 50, three different caching schemes reported 50.2, 50.3, and 49.4, respectively. Which of them is precise?

Conceptual Questions

1. When should *CoV-I* and *CoV-II* be used in the sensor-cloud?
2. What are the factors affecting the goodness of a physical node?
3. How do the models for caching compute the rate of change of environment in the sensor-cloud?
4. What are the different virtualization schemes in the sensor-cloud?
5. Why is the cache-based sensor-cloud more energy efficient than the conventional one?

Note

1 The lowest indicates the highest priority.

References

[1] S. Chatterjee and S. Misra, "Optimal Composition of a Virtual Sensor for Efficient Virtualization within Sensor-Cloud," in *Proceedings of IEEE International Conference on Communications (ICC)*, Sep. 2015, doi: 10.1109/ICC.2015.7248362.

[2] S. Chatterjee, S. Sarkar, and S. Misra, "Energy-Efficient Data Transmission Scheme in Sensor-Cloud," in *Proceedings of International Conference on Applications and Innovations in Mobile Computing (AIMoC)*, Feb. 2015.

[3] E. Ciftcioglu, A. Yener, and M. Neely, "Maximizing Quality of Information from Multiple Sensor Devices: The Exploration Vs Exploitation Tradeoff," *IEEE Journal of Selected Topics in Signal Processing*, vol. 7, no. 5, pp. 883–894, Oct. 2013.

[4] Y.-L. Lai and J. Cheng, "A Cloud-Storage RFID Location Tracking System," *IEEE Transactions on Magnetics*, vol. 50, no. 7, pp. 1–4, Jul. 2014.

[5] M. Yuriyama and T. Kushida, "Sensor-Cloud Infrastructure – Physical Sensor Management with Virtualized Sensors on Cloud Computing," in *Proceedings of the 13th International Conference on Network-Based Information Systems (NBiS)*, Sep. 2010, pp. 1–8.

[6] S. Chatterjee and S. Misra, "Dynamic and Adaptive Data Caching Mechanism for Virtualization within Sensor-Cloud," in *Proceedings of IEEE International Conference on Communications (ICC)*, Dec. 2014, doi: 10.1109/ANTS.2014.7057243.

[7] W. Liu, L. Cui, and X. Niu, "EasiTPQ: QoS-based Topology Control in Wireless Sensor Network," *Signal Processing Systems*, vol. 51, no. 2, pp. 173–181, 2008.

[8] W. Meng, L. Xie, and W. Xiao, "Optimality Analysis of Sensor-Source Geometries in Heterogeneous Sensor Networks," *IEEE Transactions on Wireless Communications*, vol. 12, no. 4, pp. 1958–1967, Apr. 2013.

[9] H. Song, L. Yu, and W.-A. Zhang, "Brief Paper – Multi-Sensor-Based H1 Estimation in Heterogeneous Sensor Networks with Stochastic Competitive Transmission and Random Sensor Failures," *IET Control Theory Applications*, vol. 8, no. 3, pp. 202–210, Feb. 2013.

[10] D.-S. Zois, M. Levorato, and U. Mitra, "Energy-Efficient, Heterogeneous Sensor Selection for Physical Activity Detection in Wireless Body Area Networks," *IEEE Transactions on Signal Processing*, vol. 61, no. 7, pp. 1581–1594, Apr. 2013.

[11] A. P. Jayasumana, Q. Han, and T. H. Illangasekare, "Virtual Sensor Networks – a Resource Efficient Approach for Concurrent Applications," in *Proceedings of the 4th International Conference on Information Technology*, 2007, pp. 111–115.

[12] R. Tynan, G. O'Hare, M. O'Grady, and C. Muldoon, "Virtual Sensor Networks: An Embedded Agent Approach, in *Proceedings of International Symposium on Parallel and Distributed Processing with Applications (ISPA)*, Dec. 2008, pp. 926–932.

[13] H. Bandara, A. Jayasumana, and T. Illangasekare, "Cluster Tree Based Self Organization of Virtual Sensor Networks," in *Proceedings of IEEE Global Communications Conference (GLOBECOM) Workshops*, Nov. 2008, pp. 1–6.

[14] S. Bhunia, J. Pal, and N. Mukherjee, "Fuzzy Assisted Event Driven Data Collection from Sensor Nodes in Sensor-Cloud Infrastructure," in *Proceedings of the 14th IEEE/ACM International Symposium on Cluster, Cloud and Grid Computing (CCGrid)*, May 2014, pp. 635–640.

[15] R. C. Ben-Yashar and S. I. Nitzan, "The Optimal Decision Rule for Fixedsize Committees in Dichotomous Choice Situations: The General Result," *International Economic Review*, vol. 38, no. 1, pp. 175–186, 1997.

Chapter 6

Pricing and Networking in the Sensor-Cloud

As sensor-cloud is an extension of the conventional cloud-computing infrastructure, it adheres to the pay-as-you-go model [1]. Conventional cloud infrastructures providing IaaS [2], PaaS [3], and SaaS [4] follow models in which end-users are charged only for the units of services they consume. In the sensor-cloud, end-users are also charged only for the services offered to them in terms of sensor data, cloud infrastructure, and resources. Therefore, the pricing strategy must quantify the resource consumption by end-users and assign reasonable charges for the usage. The payment made by the end-users is perceived at the cloud-end as profit earned for the services provided. However, a part of this profit is shared with the sensor-owners based on the usage of the sensors that are registered with the cloud [5].

In this chapter, we discuss the pricing scheme of the sensor-cloud designed specifically for provisioning Se-aaS. For conventional cloud services such as IaaS, PaaS, or SaaS, several pricing models exist today. These pricing models focus on homogeneous service-oriented architectures (SOAs) targeted towards infrastructure, platform, or software. However, the sensor-cloud is based on a heterogeneous SOA that provisions service in two distinct forms: (i) hardware and (ii) analytics and infrastructure. Thus, there is a need to design a pricing strategy specifically for Se-aaS in the sensor-cloud.

The first kind of service comes from the utilization of physical sensor devices, and the second is facilitated by the utilization of the cloud infrastructure [6]. Because of the fusion of different services involved in Se-aaS, existing pricing strategies are not fit for the sensor-cloud. Towards designing a new pricing policy, it is important to measure the consumption of every end-user and quantify the usage in terms of processing and infrastructural resources.

Requirements of end-users are likely to fluctuate heavily in practice. Thus, pricing policies within the sensor-cloud are expected to be dynamic and adaptive to changes in user demand. The goal is to optimize the price charged from the end-users and the profit margin of the cloud service providers and sensor-owners. At no point of time should end-users be overcharged; if this happens, they may terminate the service.

6.1 Scenario for Pricing

We will start by presenting the scenario for pricing within the sensor-cloud as obtained from the work of Chatterjee *et al.* [1]. The goals of the pricing strategy are the following:

i) Maximizing the profit of the CSP due to service provisioning
ii) Maximizing the profit of the individual sensor-owners depending on the usage of physical sensor nodes
iii) Ensuring that the end-users are charged reasonably (i.e., they are not overcharged and are satisfied with the services)

For every application, the CSP determines the source sensor node, which is the node responsible for sensing the data of interest to the end-user. If the source node is not directly reachable from the base station over a single hop, the CSP determines the route from the source sensor node to the destination at the cloud-end. Thus, several other nodes become part of the route, and sensor nodes are encouraged by the process of incentivization to participate in routing. The incentives can be modeled differently for systems based on profit characterization policies.

At the cloud-end, separate costs are computed for running analytics, deploying the resources (VMs, VSs, and IT resources), and maintenance of the resources. Taking into account all of these costs and their associated overhead them, the CSP formulates and computes the price to be charged the end-users. In Figure 6.1, we illustrate the overall end-to-end pricing scenario in the sensor-cloud.

6.2 The Model for Pricing

We consider the underlying sensor network, N, to be composed of m physical sensor nodes, such that $N = \{n_1, n_2, \ldots, n_m\}$. Each of these nodes is purchased by the sensor-owners (the set of all sensor-owners is represented by S) and

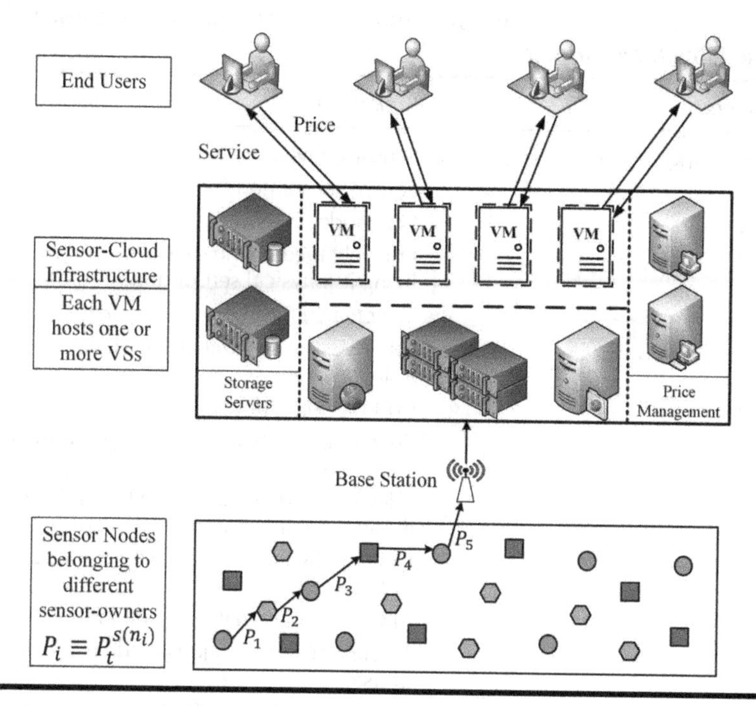

Figure 6.1 Network architecture of sensor-cloud [1]

registered with the cloud infrastructure. A sensor, n_i, is owned by the sensor-owner $s(n_i)$. The set of end-users in the system is $E = \{e_1, e_2, \ldots, e_l\}$ where each member of the set may request sensor services. The terminology of the variables is described in Table 6.1.

6.2.1 Assumptions of the Model

We now specify the underlying assumptions considered by the model [1]:

i) Only one CSP in the system collaborates with multiple sensor-owners (i.e., the system is monopolized with respect to the CSP, and oligopolized with respect to the sensor owners).
ii) The requests of an end-user are processed within a single VM, which, however, may contain one or more VSs or VSGs.
iii) It is up to the end-user to decide to continue or terminate the service at any time t. If an end-user is dissatisfied at t, s/he terminates the service at time $t + 1$.

Table 6.1 Variables and interpretation used for pricing in the sensor-cloud [1] (the table looks broken)

Parameters	*Explanation*
$S' = \{s(n_1), s(n_2), s(n_3), \ldots, s(n')\}$, $n' < n$	Sensor-owners
n_1	Source sensor node
$n_i, 2 \leq i \leq n'$	Hop node
$P_t^{s(n_j)}$, $1 \leq j \leq n'$	Price charged by the sensor-owner $s(n_j)$ for utilizing its physical sensor node at time instant t
VM_e	VM created for the end-user e, $e \in E$
$VS_e = \{vs_1, vs_2, \ldots, vs_{k(t)}\}$	Set of VSs created within VM_e at time instant t for e
$C_{VM_e}(t)$	Cost of VM_e at time t
P_{VM_e}	Price charged by the CSP from end-user e for using VM_e
$P_{vs_i}(t)$	Price charged by the CSP to the end-user e for the VS vs_i at time instant t
$\lambda^e_{vs_i}(t)$	Demand by the end-user e for the VS vs_i at time instant t
c	Criticality of the data per unit time
R	Total number of requests made by all the end-users
w	Service rate of the CSP

iv) The data and control packets are periodically transmitted from the network to the sensor-cloud-end to update the service provider about aspects related to the health of a sensor, energy content, and security threats.

v) A sensor node registered with the sensor-cloud is statically deployed. The node itself is aware of its own physical coordinates as well as of its neighbors and even the corresponding BS.

Getting deeper into Figure 6.1, we observe that the source sensor node, n_1, is owned by $s(n_1)$. n_1 senses the required information and forwards it through a multi-hop route to the cloud. In the sensor-cloud, the idea is to encourage the neighboring nodes to participate in this process of multi-hop routing. For this purpose, the nodes are incentivized by a share of the price charged to the end-user (i.e., whenever a node participates in the routing, the corresponding owner receives a share of the money paid by the end-user for the services). To begin with, n_1 selects the next intermediate node, n_2, and charges $s(n_2)$ with a price of $P_t^{s(n_1)}$. One might ask, "why would a sensor-owner charge another sensor-owner?" One can actually think of the effective price to be paid by the end-user for sensor services as the final price claimed by n'. If n_2 agrees to be the next node of the route, it accepts the price $P_t^{s(n_1)}$, and, in turn, charges $P_t^{s(n_2)}$ to n_3. So, $P_t^{s(n_2)} > P_t^{s(n_1)}$. This sequence of claiming a price continues till the last node that charges $P_t^{s(n')}$. Clearly, $P_t^{s(n')} > P_t^{s(n'-1)} > \ldots > P_t^{s(n_2)} > P_t^{s(n_1)}$. Once the data reach the cloud, the cloud resources are allocated accordingly to process and analyze the data. This includes creation of the VM and VS instances within it. For the usage of cloud resources, the service provider charges the end-user a price in addition to that charged on behalf of the sensor-owners. As per the work of Chatterjee *et al.* [1], the overall pricing scheme in the sensor-cloud has two sub-schemes:

a) Pricing attributed to Hardware (pH)
b) Pricing attributed to Infrastructure (pI).

6.3 pH: Pricing Attributed to Hardware

In this section, we present the algorithm named pH, as it concerns the pricing attributed to hardware. This typically involves the usage of sensor hardware and the pricing concerning the business with the sensor-owners. The proximity of the source sensor node n_1, in this respect, is important. If n_1 is directly within the reach of the BS by a single-hop, the data can be transmitted quickly to the BS. On the other hand, the data packet has to travel through a multi-hop route. The design of the proposed pricing model was influenced and motivated by the work of Musacchio and Walrand [7] and Lam *et al.* [8]. The pricing model ensures quality of service as well as quality of information.

6.3.1 Selection of the Next Hop Node

We will now discuss the selection procedure of the intermediate hop nodes from the source sensor node to the BS [1]. In our discussion, we limit ourselves to a single BS. However, the work can be extended to support multiple BSs and take into account issues related to resiliency, efficiency, and fault tolerance. For the purpose of selection of the next hop node, a set of nominated nodes (with respect to n_1) H_{n_1} is constructed. Let $r_{n_1}(t)$ be the communication range of n_1 at time t, and let the circular area under the direct reach of n_1 be $A_{n_1}(t) = \pi r_{n_1}(t)^2$. We consider that, within $A_{n_1}(t)$, the set of nominated hop nodes is $H_{n_1} = \{h_1, h_2, \ldots, h_b\} \mid \xi(h_j, n_1) \leq r_{n_1}(t)$, where $\xi(.)$ evaluates the inverse of the Euclidean distance between two given nodes. Every node h_i is assigned a utility value $\eta_{h_i}(t)$ at time t. The expression of $\eta_{h_i}(t)$ is shown in Definition 7. The node that has the maximum utility η_{max} is chosen as the next hop node, which is n_2.

Coming to the determination of the utility of a node h_i at time instant t, there are several contributing factors. The utility is defined as follows [1]:

Definition 7. The utility $\eta_{h_i}(t)$ of a hop node h_i, $\forall i = \{1, 2, 3, \ldots, b\}$, at time instant t, is a function of its residual energy $Q_{h_i}(t)$, its received signal strength $RSS_{h_i}(t)$, its proximity to the BS $\xi(h_i, BS)$, and its state transition overhead P_{pq}. $\eta_{h_i}(t)$ is expressed as,

$$\eta_{h_i}(t) = (Q_{h_i}(t) + g \times \frac{RSS_{h_i}(t)\xi(h_i, BS)}{P_{pq}})$$

g being a normalization factor with the same unit as that of ξ.

These utility values are computed for every node in H_{n_1} to determine the node with the maximum utility. Therefore, we have the following:

$$n_i = \max_{\forall h_k \in H_{n_i - 1}} \eta_{h_k}(t) \tag{6.1}$$

6.3.2 Context-Aware Pricing

Now that the hop node is selected, we discuss the pricing scheme, which is context-aware [1]. The price to be charged by the sensor-owner of the last hop node $s(n')$ is $P_t^{s(n')}$. The context of the data and the quality of information are considered.

Motivated by the general design for the metric "Quality of Information" (QoI) [9], the QoI ($\mathcal{Q}_{n_i}$) of node n_i at time t is formulated. If ω_{n_i} is the time-dependent discounting factor that takes into account the transmission confidence and the temporal relevance of the data, QoI is expressed as:

$$\mathcal{Q}_{n_i} = \omega_{n_i} \mathcal{Q}_{n_i - 1} + \eta_{n_i}, \mathcal{Q}_{n_1} = 1 \tag{6.2}$$

Definition 8. The price $P_t^{s(n')}$ charged by sensor-owner $s(n')$ of the last hop node n', is directly proportional to the QoI of the data of n' at time t,

$$P_t^{s(n')} \propto \mathcal{Q}_{n'}(t) \Rightarrow P_t^{s(n')} = c_1(t)\mathcal{Q}_{n'}(t) \tag{6.3}$$

where c_1 is a multiplicative factor that accounts for the parameters of signal attenuation in terms of the Nodal Signal to Noise Ratio (NSNR) [10] and the total number of transmission attempts for the corresponding packet.

Definition 9. The utility $\mathcal{U}$ of the end-user, e, is defined as the amount of data received per VS vs_i per unit time. This implies: $\mathcal{U} \sim U(\gamma_1, \gamma_2)$.

6.3.2.1 Strategy Profile

In a pricing scheme, it is crucial to have a strategy profile that enables the charging entities within the scheme to decide or reject the scheme based on the situations and the price charged. For the parties making the payment, it is important to decide whether to accept or terminate the service. Here, we present the strategy profile of the proposed pricing scheme. The design of the strategy profile is motivated by the works of Lam *et al.* [8], and Fudenberg and Tirole [11]. As obtained from the work of Chatterjee *et al.* [1], we state the strategy profile below:

- The end-user e obtains data from a VS, vs_i, for a time period, τ. The end-user follows a myopic strategy: it retains a VS, vs_i, at time, t, if $(t \leq \tau)$ and $(\mathcal{U} \geq p_t^{s(n')})$ i.e., within the time period, τ, the end-user accepts the service if and only if the utility, $\mathcal{U}$, is higher than the price to be paid by the end-user.
- The sensor-owner, $s(n_i)$, $\forall i = \{2, 3, \ldots, n'\}$, of a participating hop node charges a price, $p^{*s(n_i)}(p^{s(n_{i-1})})$, which is dependent on the price charged by the previous sensor-owner, $s(n_{i-1})$ [1].
- The owner of the last hop node, $s(n')$, charges a decreasing price sequence, $\{p_t^{s(n')}\}$.

With the assumption that γ_1, γ_2 are known, the optimal price, $p^{*s(n')} \in [\gamma_1, \gamma_2]$, charged by. $s(n')$, is determined. Further, with respect to $p^{*s(n')}$, the optimal price, $p^{*s(n_i)}$, charged by the owner, $s(n_i)$, of any participating hop node, $\forall i = \{2, 3, \ldots, n'\}$, can be obtained as well.

6.4 pI: Pricing Attributed to Infrastructure

Now we discuss the pricing attributed to infrastructure, – pI [1]. When an end-user, e, requests Se-aaS, the service provider creates a VM specifically dedicated to the end-user, VM_e. As mentioned before, the number of VMs is considered to be one for every user, in our case. Based on the requirements of the application, at any time, t, $k(t)$ number of VSs are created and stored within VM_e. Every VS has a demand that is essentially the information rate expected from the VS per unit time. Based on $\lambda_{vs_i}^e(t)$, the service provider charges $P_{vs_i}(t)$ at time t. The service provider incurs a cost of $C_{VM_e}(t)$ for creating VM_e. This cost is the summation of spawning the instance of the VM, B_{VM_e}, and the maintenance cost of the VM from the time it is alive, t_{built}, till the time it is killed. Therefore, we have:

$$C_{VM_e}(t) = B_{VM_e}(t) + M_{VM_e}(t - t_{built}). \tag{6.4}$$

The maintenance cost of a VM, M_{VM_e}, is the summation of the cost for creating every VS, $vs_i \in VS_e$, and maintaining it per unit time. Thus,

$$M_{VM_e}(t - t_{built}) = \sum_{i=1}^{k(t)} (B_{vs_i} + M_{vs_i}(t - t_{0i})) \tag{6.5}$$

t_{0i} is the time of creation of vs_i. Therefore, the overall cost incurred by the CSP for provisioning infrastructural resources to end-user e at time t is expressed as

$$C_{VM_e}(t) = B_{VM_e}(t) + \sum_{i=1}^{k(t)} B_{vs_i} + \sum_{i=1}^{k(t)} M_{vs_i}(t - t_{0i}). \tag{6.6}$$

Generally, a VS contains sensors with homogeneous sensing hardware to serve an application. The control for creating, updating, or deleting a VS is at the will of the end-user. However, there is a maintenance cost associated with the VS because of which end-users might want to delete them if they will not be usable for a significant duration of time.

Definition 10. The net profit of the CSP at time, $t, r(t)$, is defined as the difference of the total price charged the end-user and the sum of the cost incurred in creating and maintaining the VM for a particular end-user, e, and the overall price charged through pH for $e\,(p_e^{s(n')}(t))$. Thus, $r(t)$ is expressed as,

$$r(t) = \left(\sum_{i=1}^{k(t)} \lambda_{vsi}^{e}\,(t) P_{vs}(t) \right) + P_{VM_e} - C_{VM_e}(t) - p_e^{s(n')}(t), \tag{6.7}$$

where the price charged for each VS is a function of the rate of change of demand for each $vs_i(t)$.

Definition 11. The user satisfaction, $u_e(t)$, for a particular end-user, e, at any time instant, t, is a function of the total demand made by e for all the VSs within VM_e, the total cost incurred at the sensor-cloud end for serving the demand, and the total price charged by the CSP [1].

The overall profit of the service provider is considered over time t taking into account the user satisfaction. We have:

$$\mathcal{F}(T) = \sum_{t=0}^{T} r(t) \qquad 6.8)$$

subjected to some constraints [1]. $\mathcal{F}$ is maximized using the approach of Lagrangian Multiplier.

6.5 Real-Life Applicability: A Case Study

In this section, we present a real-life scenario of pricing in the sensor-cloud [1]. We demonstrate the scenario with the example of an environment monitoring application as shown in Figure 6.2. The figure illustrates the sequence of different activities from the request of a service to eventually paying for the service.

It all starts with the request from the end-user to obtain Se-aaS corresponding to a environment monitoring application (1). The user specifies the requirement through high-level templates. Following this, the cloud servers decode the requests in terms of compatible environmental sensors (2). The chosen sensors are activated and scheduled. At this stage, the process of virtualization commences when the data from multiple sensors start logging in (3) [2,12,13]. To store and process data from the VSs, VMs are initialized and set up for the user. At this moment, the pricing scheme begins taking into account the costs due to the initialization of the VM (4). Once the entire setup is arranged (5), the VM is accessed and the data processing starts (6). The pricing unit transmits the sequence of prices charged over time (7) to the end-user for the services offered. Simultaneously, the quantification of the usage also takes place (8). The algorithms of pH (9) and pI (10) are invoked respectively. The payments of the end-users are eventually channelized to the service provider and the different end-users (11).

6.6 Networking

As we know, Se-aaS is provisioned in the form of VSs [5,6]. The VSs may be composed of multiple physical sensors from single or multiple geographical locations. As the data from these geographically scattered VSs are deposited to the nearest data centers (DCs), clearly, geo-distributed DCs are involved in provisioning Se-aaS to end-users.

For the sake of data analysis and processing, VSs serving an application must co-exist within a single VM in a particular DC. However, as the VSs could be geographically scattered, there is a requirement to migrate the VSs or the content

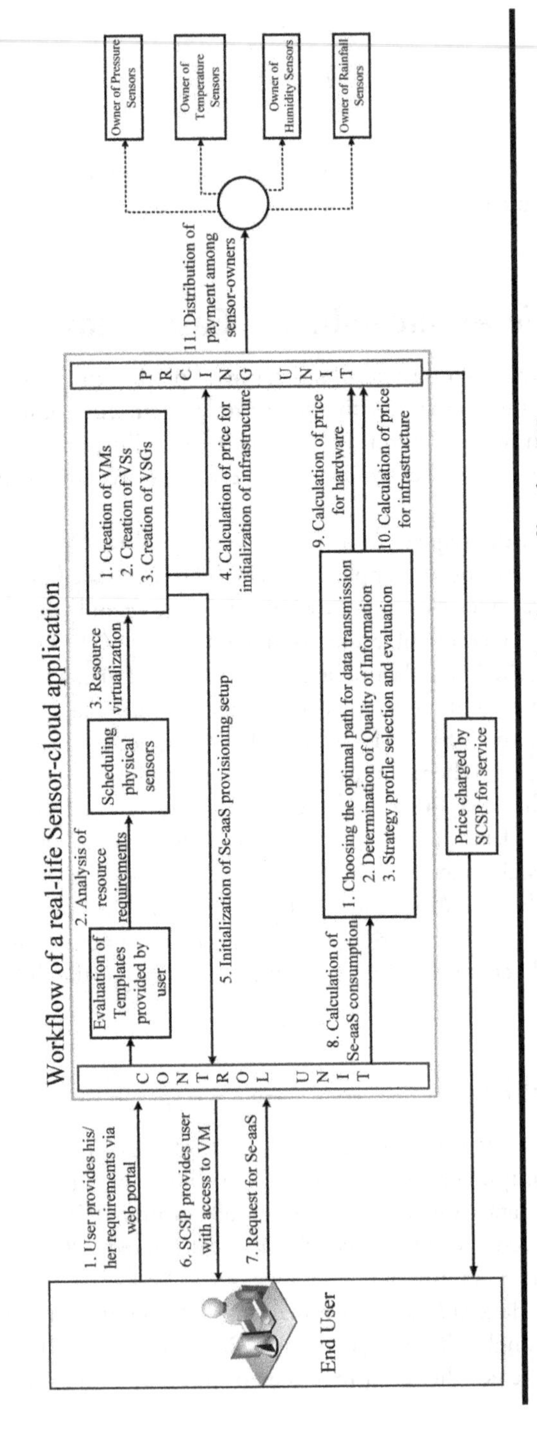

Figure 6.2 The applicability of pricing within an environment monitoring application [1]

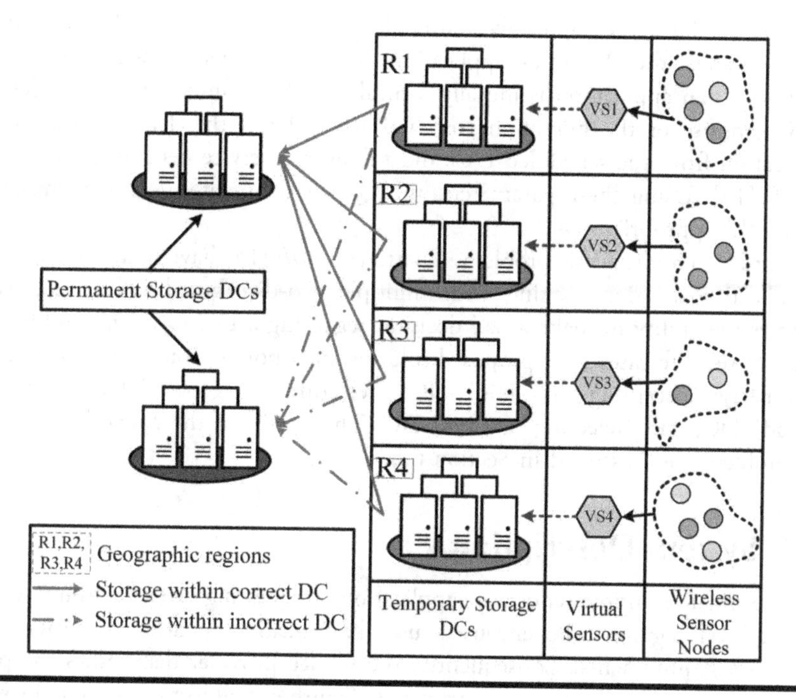

Figure 6.3 Different storage types in DCs [14]

of the VSs to a single VM that serves as the center of data processing for a particular end-user. Therefore, it is extremely important to design and implement a dynamic DC selection algorithm within the sensor-cloud for every end-user application.

Figure 6.3 shows n geographically scattered non-overlapping regions that contain physical sensor nodes that serve an application [14]. Every region contains separate physical nodes that make up the VS. Each physical sensor transmits its data to the nearest DC of the corresponding region. Such DCs are called temporary storage DCs. It is required to gather the data from multiple DCs to a single one for further analytics and processing. The finally selected DC contains the VM serving the end-user. In this regard, the question is how to select the DC that actually serves the end-user through its VM. This is particularly challenging as different DCs have different capacities with respect to resource, application, and decision-making abilities. Therefore, we cannot simply go with a random selection of a DC, as it leads to network overhead and reduction of QoS of application. This is where the problem of optimal selection of a DC is crucial with respect to minimizing the network overhead and providing the maximum QoS.

With the DCs geographically scattered, the time for data transmission from DCs will affect the QoS of an application. In the work that we will now discuss, the QoS of an application is initially formulated. The components to model the QoS comprise of the migration cost within the DCs, the delivery cost to the application from the scheduled DC, and the overall service delay of provisioning Se-aaS [14]. Using these parameters, the QoS is formulated and maximized to choose the appropriate DC.

In order to address this problem, Chatterjee *et al.* [14] have used the approach of collective decision making with multiple geo-distributed DCs. The work assumes the ability to make a bad decision regarding a DC to fit the problem to practical real-life cases. The proposed solution incorporates four different types of asymmetry: "accepting" a "good" DC, "rejecting" a "good" DC, "accepting" a "bad" DC, and "rejecting" a "bad" DC. The details of the characterization of the problem are illustrated in Section 6.8.

6.7 System Description

We consider multiple end-user applications requesting Se-aaS from several geographical regions. The end-users use the Se-aaS templates to specify their high-level requirements. Consequently, the service provider determines the physical sensors to be allocated and activated. Figure 6.4 illustrates a scenario with different non-overlapping geographical regions, R_i [14]. Physical sensors from a particular region are logically grouped to form a VS. Multiple such VSs serve an application. The data from each VS are deposited into the nearest DC of that region. However, based on the working principle of the sensor-cloud, for further analytics and aggregation, the data have to travel to a single VM within a single DC [6,12]. Thus, it is important to choose a DC that will contain the VM for data aggregation.

The final DC selected for an application contains the VMs to store and process sensor data. The data from the VSs are migrated from the respective geo-distributed DCs to a single DC. The selection of DC heavily controls the QoS of the application, which depends on the service delay of the end-user, as well as the migration cost of the DCs.

6.8 Formal Definition of the Problem

The system comprises a set $\mathcal{D}$ of ω DCs, such that $\mathcal{D} = \mathcal{D}_1, \mathcal{D}_2, \ldots, \mathcal{D}_\omega. A\ DC\, \mathcal{D}_i$ is positioned at region R_i. Hence, the set of regions is $\{R_1, R_2, \ldots, R_\omega\}$, as indicated in Figure 6.5. Let us assume that m types of sensor services are requested by an application, App_i. Thus, the corresponding VSs are represented as $\mathcal{V}_1, \mathcal{V}_2, \ldots, \mathcal{V}_m$ and are dumped in temporary DCs, $\mathcal{D}_1, \mathcal{D}_2, \ldots, \mathcal{D}_m$, $(m \leq \omega)$, respectively. These

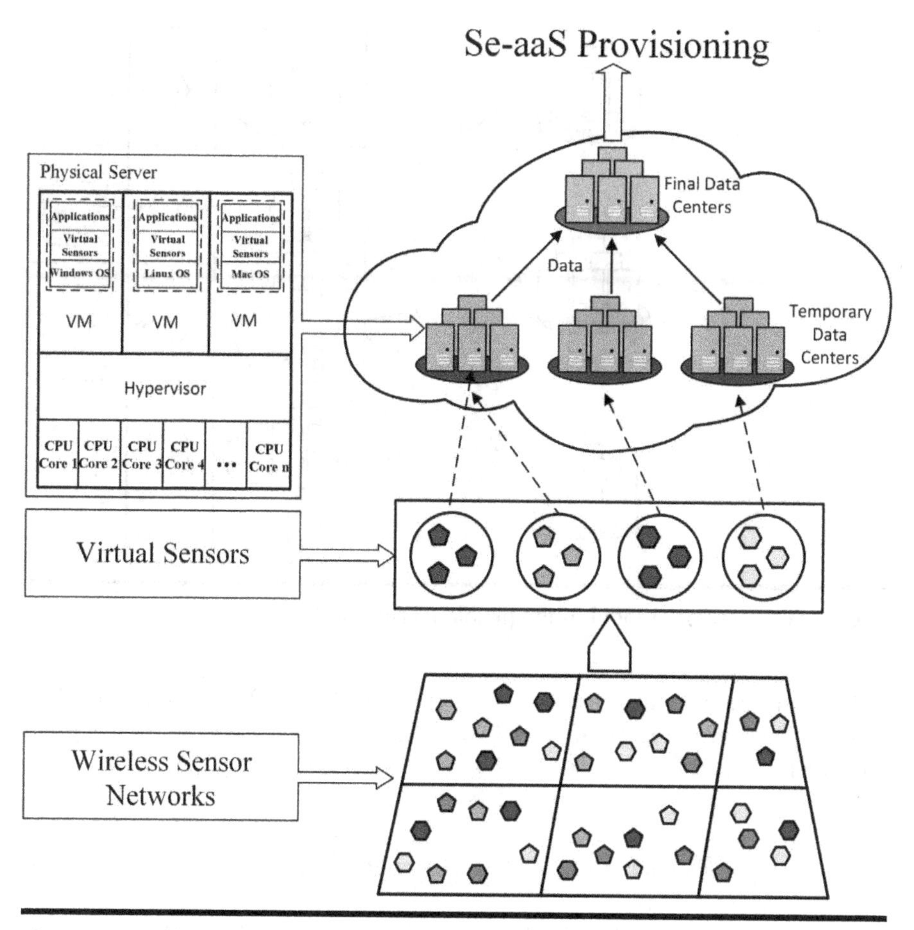

Figure 6.4 Diagrammatic representation of the system [14]

temporary DCs capture the released sensor data from the nodes. The data then travel to the final DC of the application. Thus, the goal is to select a DC, $\mathcal{D}_w^{App_i}$, for App_i by maximizing QoS.

Definition 12. The migration cost of sensor data from DC, $\mathcal{D}_i$, to DC, $\mathcal{D}_j$, for transmitting p packets is denoted by $\mathcal{M}(\mathcal{D}_i, \mathcal{D}_j)$ and is defined in terms of the latency, $\mathcal{L}$, involved in delivering a packet from $\mathcal{D}_i$ to $\mathcal{D}_j$. The function, $\mathcal{M}(\cdot, \cdot)$, is expressed as $\mathcal{M}(\mathcal{D}_i, \mathcal{D}_j) = \mathcal{L}(\mathcal{D}_i, \mathcal{D}_j, p)$, where $\mathcal{L}(\mathcal{D}_i, \mathcal{D}_j, p)$ is the latency involved in migrating p packets, each of size P byte, from $\mathcal{D}_i$ to $\mathcal{D}_j$.

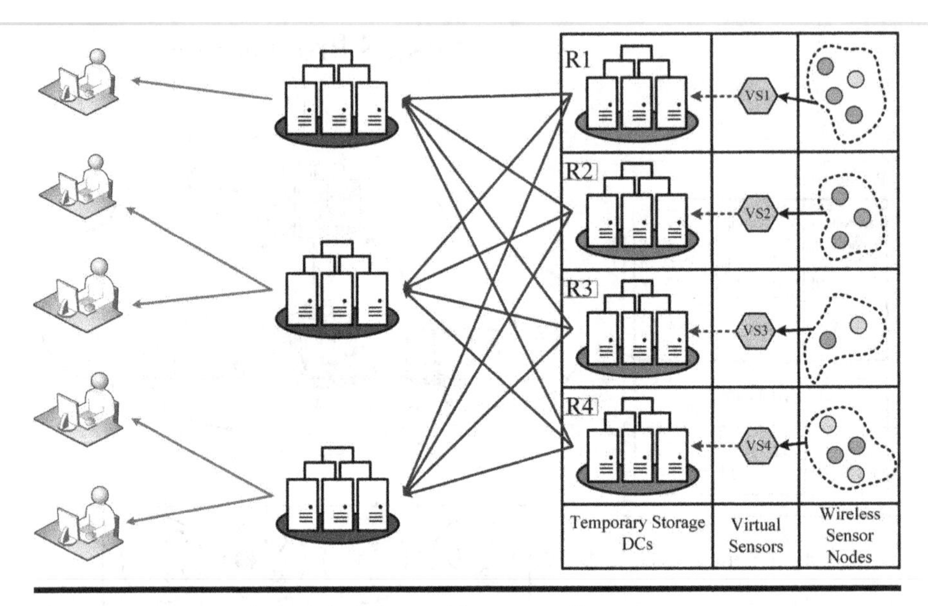

Figure 6.5 Network model of the problem scenario [14]

Definition 13. The delivery cost of p packets from $\mathcal{D}_i$ to an application, App_j, at absolute location, $(l_{1,j}, l_{2,j})$, is denoted as $\delta(\mathcal{D}_i, l_{1,j}, l_{2,j})$ and is expressed as:

$$\delta(\mathcal{D}_i, l_{1,i}, l_{2,i}) = (\sqrt{(\mathcal{D}_i.x - l_{1,i})^2 + (\mathcal{D}_i.y - l_{2,i})^2})/\eta, \tag{6.9}$$

where η is in meter per second through the link connecting a DC to an end-user. Therefore, δ is finally expressed in seconds.

Definition 14. The service delay of App_i, $\mathcal{S}(App_i)$, is the summation of its migration cost and the delivery cost. It is defined as:

$$\mathcal{S}(App_i) = \sum_{i=1}^{m} \mathcal{M}(\mathcal{D}_i, \mathcal{D}_s) + \delta(\mathcal{D}_s, l_{1,i}, l_{2,i}), \tag{6.10}$$

where the data of the VSs are migrated to $\mathcal{D}_s$ from multiple DCs, $\{\mathcal{D}_i\}, 1 \leq i \leq m$.

Having defined the migration cost, the delivery cost, and the service delay, we now illustrate the formation of the QoS offered (in byte per second) for transmission of p packets to an application, App_i.

$$\mathcal{Q}_{net}(App_i) = \frac{pP}{[\sum_{i=1}^{m} \mathcal{M}(\mathcal{D}_i, \mathcal{D}_s) + \delta(\mathcal{D}_s, l_{1,i}, l_{2,i})]}. \tag{6.11}$$

Therefore, the objective function can be stated as: $f : \mathcal{D}^m \rightarrow \mathcal{D}, \mathcal{D} = \{\mathcal{D}_i\}, 1 \leq i \leq m \leq d$, which is essentially a mapping from a set of DCs to the final DC $\mathcal{D}_w$. The goal of f is to maximize the QoS,$\mathcal{Q}_{net}$; f belongs to the solution set of the maximization function:

$$\arg\max(\mathcal{Q}_{net}(App_i)), \tag{6.12}$$

For the sake of better understanding, the readers may note that the DC selection algorithm can be independently applied to several sensor-based applications such as environmental monitoring, surveillance applications, multimedia applications, traffic monitoring applications, and so on. Quite clearly, any application will have its own sensor hardware and configuration demands. Sensor-cloud is designed to take care of the issues related to inter-operability and compatibility [1,2,13,15].

The solution to the problem is based on *Optimal Decision Rule* [16] that considers the "general pairwise choice framework." In such a framework, fallibility or the possibility of making errors is considered. The rule is mainly used within committees that require a decision to be made especially in domains of investment projects, share markets, legal organizations, and economic organizations. It is based on four different types of asymmetry. Thus, in an n-member committee, there is the possibility of making mistakes by accepting bad projects and rejecting good ones. The possibilities of Type-I and II errors are taken into account to eventually reach at the decision of the committee. The decision rule maximizes the payoff in terms of the profit.

6.9 Complexity Analysis

We now analyze the complexity of the algorithm for selection of the DCs to justify its suitability for real-life applications or real-time processing. In Figure 6.6, the runtime complexity of the algorithms is demonstrated by the metric of simulation time, as recorded during experimentation. In the first experiment, the number of applications n is set to 100. The number of DCs serving the applications is varied.

In Figure 6.6(a), we observe that the number of temporary DCs is varied from 10 to 100. The number of nominated DCs is set to a fixed value, $\mathcal{D}^{nom} = 9$. The algorithm was executed and iterated 50 times for each value set for temporary DC. The final result was plotted within a 95% confidence interval. We observe that the mean time to simulate or the complexity of the algorithm is 0.024 second within the boundary of $[0.022, 0.026]$ second.

In Figure 6.6(b), the number of nominated DCs was increased from 5 to 15, whereas the number of temporary DCs was fixed, $\mathcal{D}_{App_i} = 50$. The runtime of the simulations was observed to lie within the 95% interval of $[0.0214, 0.0238]$ second, whereas the mean was observed at 0.0226 second. The analyses of the obtained results highlight that the mean variance of complexity with the increase in the number of temporary DCs is 8.9×10^{-6} second and with the increase in the number of nominated DCs is 5×10^{-6} second. Therefore, the increase in the complexity of the algorithm with the increase in data volume is negligible for real-life scenarios.

6.10 Summary

In this chapter, we discussed the topic of pricing and networking in the sensor-cloud. We presented a model for pricing in the sensor-cloud and discussed the limitations of the existing models of cloud-based pricing. Within the model, we discussed the two important components of pricing: pricing due to hardware and pricing due to infrastructure. We also demonstrated the real-life applicability of pricing with the example of an environment monitoring application. Next, we discussed data center networking in the sensor-cloud. We focused on the problem of the selection of a data center when an end-user is interested in obtaining Se-aaS from geographically distributed sensor nodes. We presented the formal definition of the problem and the complexity analysis of such solutions.

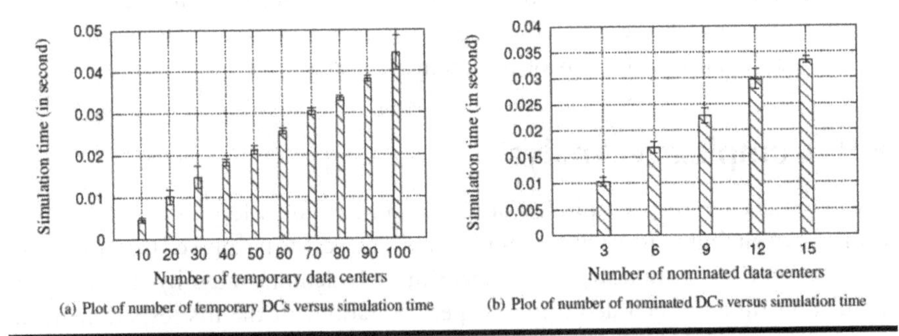

Figure 6.6 Complexity analysis by varying system components [14]

Working Exercises

Numerical Problems

1. A node is 2 m away from the BS with an energy content of 1 nJ. The RSS of the node is 1.2 with a state transition overhead of 2. What is the utility of the node?
2. The cost for creating a VM in the sensor-cloud is Rs. 1000. The VM comprises 20 VSs, each of which is built and maintained at Rs. 10 and Rs. 5 per second, respectively. What is the cost incurred at the cloud-end for 5 seconds?
3. An end-user utility is uniformly distributed within [10, 100], and another end-user utility is uniformly distributed within [50, 70]. Who has the higher chance of a greater utility?
4. What is the QoS for transferring 10 2-byte packets if the migration distance is 10 ms and the distance between the data center and the application is covered in 100 ms?
5. A service provider serves 3 end-users per second. Using Poisson distribution, what is the probability that in a given day, he will serve some end-users?

Conceptual Questions

1. What are the basic assumptions of pricing in the sensor-cloud?
2. What is the myopic strategy of the end-user in pricing?
3. In case of pH, how are next hop nodes selected?
4. Why is context-aware pricing important in the sensor-cloud?
5. What is the basis of the optimal decision rule?

References

[1] S. Chatterjee, R. Ladia, and S. Sarkar, "Dynamic Optimal Pricing for Heterogeneous Service-Oriented Architecture of Sensor-Cloud Infrastructure," *IEEE Transactions on Services Computing*, Jul. 2015, doi: 10.1109/TSC.2015.2453958.

[2] Y. Feng, B. Li, and B. Li, "Price Competition in an Oligopoly Market with Multiple IaaS Cloud Providers," *IEEE Transactions on Computers*, vol. 63, no. 1, pp. 59–73, 2014.

[3] H. Lu, X. Wu, W. Zhang, and J. Liu, "Optimal Pricing of Multi-Model Hybrid System for PaaS Cloud Computing," in *Proceedings of International Conference on Cloud and Service Computing (CSC)*, Nov. 2012, pp. 227–231.

[4] D. Ardagna, B. Panicucci, and M. Passacantando, "Generalized Nash Equilibria for the Service Provisioning Problem in Cloud Systems," *IEEE Transactions on Services Computing*, vol. 6, no. 4, pp. 429–442, 2013.

[5] M. Yuriyama and T. Kushida, "Sensor-Cloud Infrastructure – Physical Sensor Management with Virtualized Sensors on Cloud Computing," in *Proceedings of the 13th International Conference on Network-Based Information Systems (NBiS)*, Sept. 2010, pp. 1–8.

[6] S. Madria, V. Kumar, and R. Dalvi, "Sensor Cloud: A Cloud of Virtual Sensors," *IEEE Software*, vol. 31, no. 2, pp. 70–77, Mar. 2014.

[7] J. Musacchio and J. Walrand, "Wifi Access Point Pricing as a Dynamic Game," *IEEE/ACM Transactions on Networking*, vol. 2, pp. 289–301, 2006.

[8] R. K. Lam, D. M. Chiu, and J. C. Lui, "On the Access Pricing and Network Scaling Issues of Wireless Mesh Networks," *IEEE Transactions on Computers*, vol. 56, no. 11, pp. 1456–1469, 2007.

[9] E. Ciftcioglu, A. Yener, and M. Neely, "Maximizing Quality of Information from Multiple Sensor Devices: The Exploration Vs Exploitation Tradeoff," *IEEE Journal on Selected Topics in Signal Processing*, vol. 7, no. 5, pp. 883–894, Oct. 2013.

[10] S. Sheng and R. Gao, "Structural Dynamics-Based Sensor Placement Strategy for High Quality Sensing," *IEEE Sensors, 2004*, vol. 2, pp. 642–645, Oct. 2004.

[11] D. Fudenberg and J. Tirole, Game Theory, Chapter 8. MIT Press, 1991.

[12] S. Chatterjee and S. Misra, "Target Tracking Using Sensor-Cloud: Sensor-Target Mapping in Presence of Overlapping Coverage," *IEEE Communications Letters*, vol. 18, no. 8, pp. 1435–1438, Aug. 2014.

[13] S. Chatterjee and S. Misra, "Dynamic and Adaptive Data Caching Mechanism for Virtualization within Sensor-Cloud," in *Proceedings of IEEE International Conference on Communications (ICC)*, Dec. 2014, doi: 10.1109/ANTS.2014.7057243.

[14] S. Chatterjee, R. Misra, and S. U. Khan, "Optimal Data Center Scheduling for Quality of Service Management in Sensor-Cloud," *IEEE Transactions on Cloud Computing*, Vol. 7, pp. 89–101, 2015.

[15] S. Chatterjee and S. Misra, "Optimal Composition of a Virtual Sensor for Efficient Virtualization within Sensor-Cloud," in *Proceedings of IEEE International Conference on Communications (ICC)*, Sept. 2015, doi: 10.1109/ICC.2015.7248362.

[16] R. C. Ben-Yashar and S. I. Nitzan, "The Optimal Decision Rule for Fixed-Size Committees in Dichotomous Choice Situations: The General Result," *International Economic Review*, vol. 38, no. 1, pp. 175–186, 1997.

Chapter 7

Sensor-Cloud for Internet of Things

7.1 Introduction

To recapitulate, the Internet of Things (IoT) refers to the billions of physical devices that are spread across the world and are interconnected over the Internet. Every "thing" in the IoT refers to an object that is empowered with its capacity to sense and transmit data over the Internet. A report by Forbes defines the IoT as follows [1].

> It is the concept of basically connecting any device with an on and off switch to the Internet (and/or to each other). This includes everything from cellphones, coffee makers, washing machines, headphones, lamps, wearable devices and almost anything else you can think of. This also applies to components of machines, for example a jet engine of an airplane or the drill of an oil rig.

The vision of IoT is connecting people and things alike across the globe. However, the question is, how can such a paradigm be realized? As mentioned in Chapter 3, the IoT has the characteristics of heterogeneity, unique device ID, Internet connectivity, and ubiquity [2].

To fulfill these properties, MEMS-based sensors [3–5] are the key components of the billions of Internet-connected devices. These so-called objects are powerful enough to sense the surroundings and transmit the sensed data periodically. However, there are certain challenges involved in designing and developing these sensors [6].

1. **Technology**: Building such sensor-based devices involves constraints in design, integration, heterogeneous hardware, miniaturized architecture, energy efficiency, and performance. In order to address these challenges, vendors are expected to incorporate intelligence within these devices. This, in turn, calls for leveraging the technology across multiple platforms.
2. **Cross-platform solutions**: One of the primary markets of MEMS-based sensors is controlled by the smartphone [1,7–9]. Sensor specifications are decided and set by the manufacturers of smartphones. However, for the IoT ecosystem, the specifications of the sensor, controller, and actuator vary widely based on the type of products. Thus, it is challenging and essential for vendors to incorporate cross-platform solutions to customize requirements for different applications.
3. **System complexity**: The system complexity in designing and developing sensor-based "things" is multi-fold. We expect solutions to reduce complexity and be more energy efficient, flexible, and cost-effective. As a single company or organization cannot provide or meet with all requirements, it is imperative to collaborate and cooperate when building IoT products [2,10–12].

7.2 Enabling IoT through Sensor-Cloud

As mentioned in Chapter 4, sensor-cloud infrastructure interfaces between the physical and cyber world to render Se-aaS. The primary advantage of sensor-cloud's not owning a sensor but still being able to use it, is similar to what the IoT envisions for its customers, as shown in Figure 7.1. In the IoT, in addition to sensors for personal use, as in vehicles, household appliances, laptops, and smartphones, customers have access to and the benefits of sensors deployed to meet public demand for items like cameras deployed across streets; monitors on bridges to track the health and velocity of vehicles; motion detection sensors deployed for safety purposes in airports, ATMs, and other public places, and so on. Sensor-cloud has already taken its first step towards the realization of the IoT's vision by treating the sensors as not merely hardware but as service modules. In this chapter, we depict how the different aspects of sensor-cloud contribute to the conceptualization of IoT.

7.2.1 Contributions through Architecture

As sensor-cloud is an extension of the conventional cloud-computing paradigm, end-users of sensor cloud are not aware of the complex back-end processing and analytics. Whenever an end-user requests sensor services, the cloud infrastructure provides them in terms of a virtual sensor (VS), as explained in Chapter 6. This

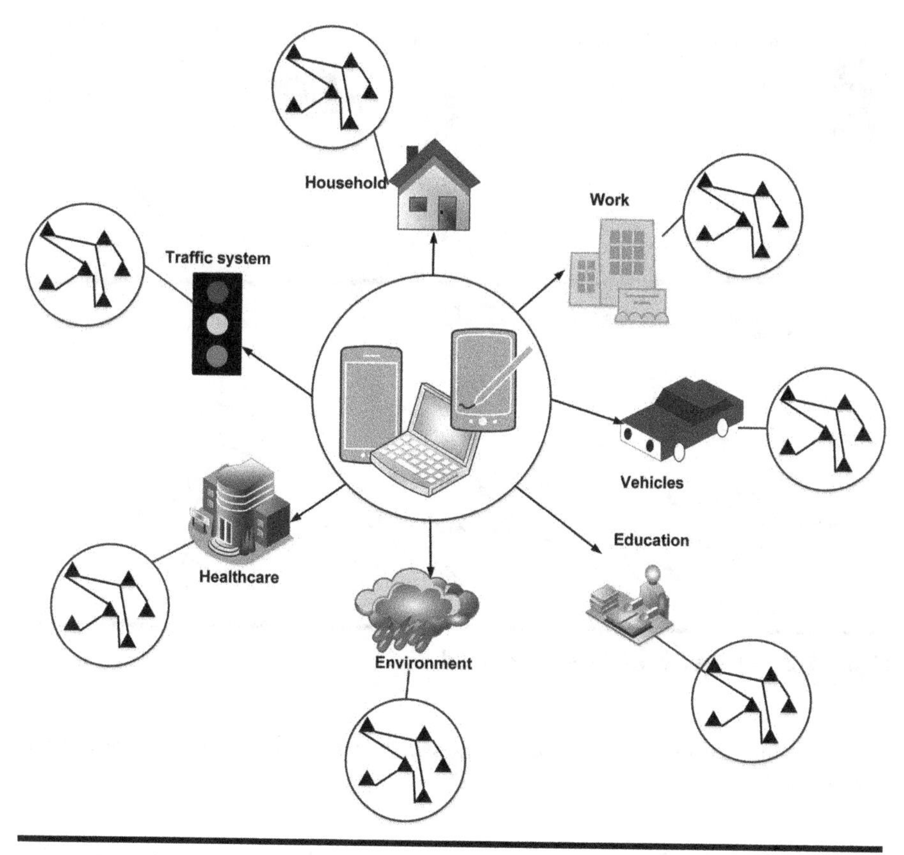

Figure 7.1 Overview of IoT

directly contributes to physical sensor management in the IoT. These physical sensors are supervised, managed, and coordinated in real time by the sensor-cloud administrator. Sensor-cloud subsidizes the sensor management aspects, which, in turn, automatically ensure improved data management in the IoT, as indicated in Figure 7.2.

Virtualization: The architecture of sensor-cloud contributes to the realization of the IoT through its inherent characteristics of virtualization that takes into account that no application has the requirement of using all kinds of sensors at all times [13–15]. Therefore, sensor-cloud is purely based on strategic sensor grouping and allocation to serve applications in real time. Data sharing and services are managed to the extent that an end-user feels that s/he has a dedicated sensor to meet his/her requirements at all times. If we delve into the advantages of the IoT, one of the primary concerns is to

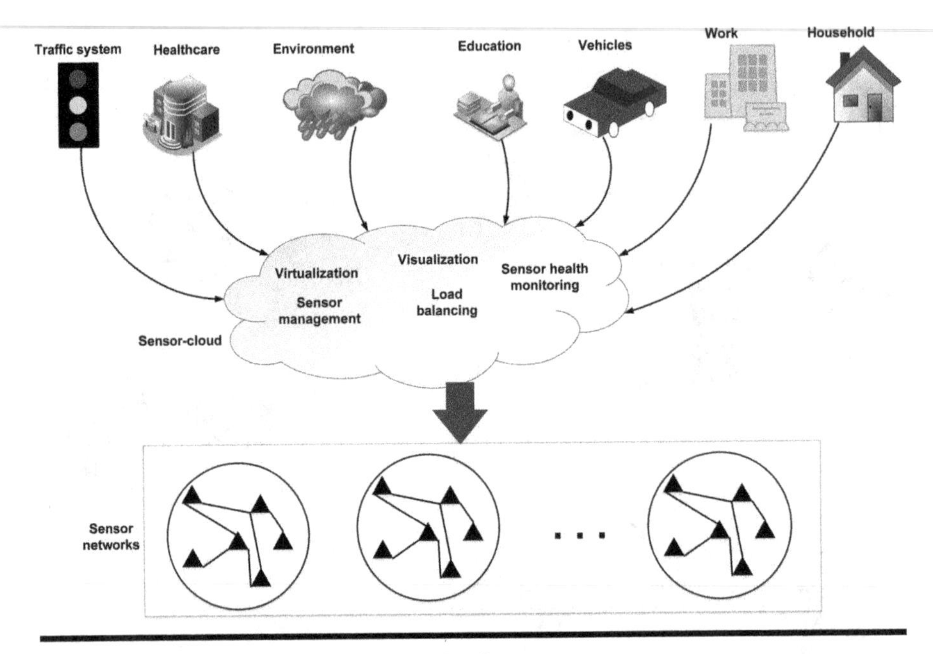

Figure 7.2 Overview of the sensor-cloud

ensure all-time sensor services to customers, which is aligned with the offerings of the sensor-cloud.

Publish/Subscribe: Services to applications are backed up and supported by the publish/subscribe mechanism used in sensor-cloud. Based on the application requirements, multiple sensor networks publish sensed information that can be subscribed to on demand. Subscription to sensor networks can be dynamic, if the required sensed information has already been published by other sensor networks. Otherwise, it is important to select the best sensors (or networks) for publishing sensed information to meet the requirement of the application.

Sensor health monitoring: Based on the requirements of an application, VSs are created and allocated to end-users. Before grouping physical sensors to the VSs, it is important to check the status of the sensor in terms of its health, energy content, and availability. This aspect is crucial from an IoT perspective. Monitoring the health and fault rates of the sensor networks is indispensable, as these networks form the backbone of the IoT. Therefore, this contribution of the sensor-cloud is another step towards the realization of the IoT.

7.2.2 Contributions to the IoT

In this section, we discuss the defining characteristics of sensor-cloud and highlight their implications for the IoT.

Analysis: Sensor data analytics is a key component for both sensor-cloud and the IoT. The sensor-cloud has already paved a way for significant advancement, as it supports sensor-based applications with customized query processing and runtime analytics. These analytics are supported by the scale of the infrastructure. Therefore, on-demand scalability of resources is another contribution of sensor-cloud to the IoT. It is important to mention here that the IoT supports billions of users with heavy workload fluctuations. Thus, the support of resource scalability at application runtime can be considered a vital footprint.

Collaboration: The, the manufacturing of IoT devices enforces and encourages cross-platform design and integration. In this regard, sensor-cloud is still emerging; the focus is on integrating heavily heterogeneous sensors in hardware, communication, and sensing range, communication protocols, and standards. However, ongoing research is a driving factor of collaborative sensing in IoT.

Visualization: Sensor-cloud provides an Application Programming Interface (API) to the end-users for managing sensors on their own. This includes not only launching customized queries from the API, but also facilitating simple sensor operations such as adding or deleting a sensor or selecting a particular sensor for the current services. This is relevant from an IoT perspective, as the IoT envisions provisioning an interface that helps customers visualize the sensor representation. These interfaces could be present anywhere: smartphones, laptops, and other digital assistants. Using these interfaces, customers can issue queries (e.g., an end-user may issue a query from his/her workplace to enquire if the refrigerator is running). Sensor-cloud has made significant improvement towards sensor visualization aspects; they will contribute immensely for sensor visualization in the IoT.

7.2.3 Contributions through the Life Cycle

In this section, we describe how the life cycle of the sensor-cloud is compatible with the life cycle model of the IoT. We start by introducing the life cycle of the IoT and then deduce and correlate the strings of coherence with that of sensor-cloud architecture.

According to RedHat [16], as shown in Figure 7.3, the IoT information life cycle comprises four distinct phases: (i) data, (ii) information, (iii) intelligence,

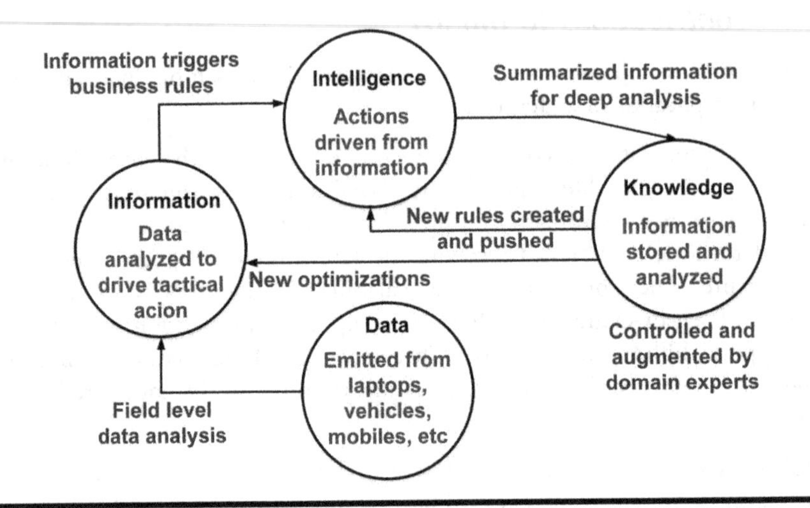

Figure 7.3 Life cycle representation of the IoT [16]

and (iv) knowledge. The first phase (*data*) involves the collection of raw sensed data generated from devices or "things" such as smartphones, laptops, and vehicles. These data undergo a field level analysis before the commencement of the second phase (*information*). In the second phase, data are analyzed by application requirements to determine the actions to be taken. Then begins the third phase (*intelligence*), in which information is analyzed to adapt to customer needs. In the last phase (*knowledge*), the acquired intelligence undergoes deeper analysis, and it is added to the database of intelligence and the tool for data analysis.

The life cycle of the sensor-cloud consists of the following stages: (i) preparing resources and sensor services, (ii) preparing templates for Se-aaS, (iii) accepting user requests, (iv) providing the services, (v) unregistering the service instances, and (vi) unregistering Se-aaS from an end-user, as indicated in Figure 7.4 [17]. In the first stage, the sensor-cloud allocates physical resources to serve end-users. These include the physical storage, CPU cores, memory, and network resources. This stage also includes checking and ensuring correct deployment of physical sensors across several geographical regions. In the second phase, templates are created using Sensor Modeling language (SML) to represent the sensor metadata using XML representations. The idea of using such representation is to facilitate the use of templates across multiple platforms, hardware, and operating systems. In the third phase, the end-users request services through the templates in a high-level fashion. In the fourth phase, services are decoded within the cloud infrastructure in terms of physical sensor allocation. These services are inclusive

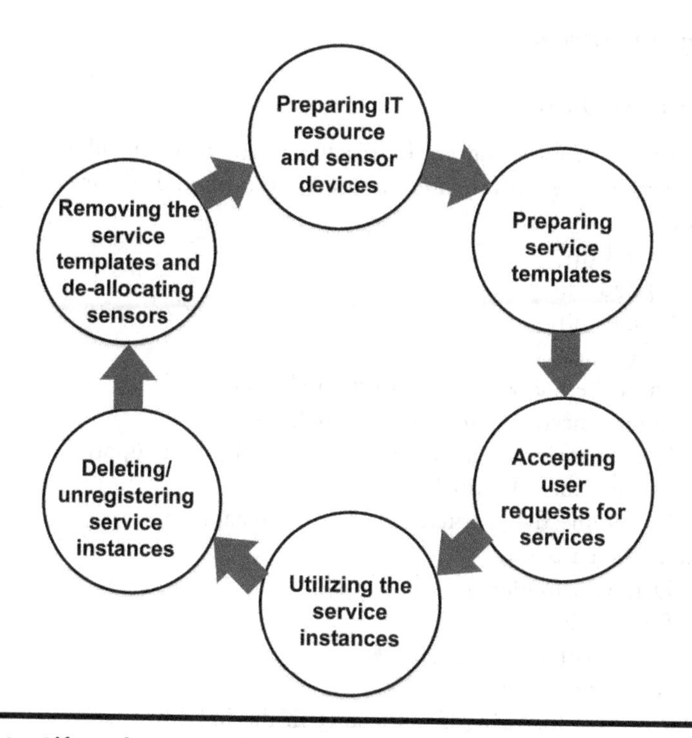

Figure 7.4 Life cycle representation of the sensor-cloud [17]

of sensor scheduling, load balancing, and virtualization. The fifth phase commences when the end-user has been served and service instances are deleted. This includes release of resources serving the end-users – CPU cores, memory, and VMs. Finally, in the sixth phase, the physical sensors are released to make them available to serve the next users.

The above stages clearly demonstrate the implications of sensor-cloud for the IoT. Sensor-cloud paved the way through its functionalities and models for future sensor-based technology. The architecture of sensor-cloud is aligned with that of the IoT, which makes some of the components and modules re-usable with little or no modification.

7.3 Summary

In this chapter, we learned the implications of the sensor-cloud on the IoT. We primarily focused on how sensor-cloud infrastructure enables the realization of the IoT through its contributions in terms of architecture, functionalities, and the life cycle of operations.

Working Exercises

Multiple Choice Questions

1. Which of the following are IoT features – (i) connect billions of devices, (ii) every device has a communication protocol, and (iii) every device can sense?
   a) (i) and (ii)
   b) (ii) and (iii)
   c) (i) and (iii)
   d) (i), (ii), and (iii)
2. Which of the following are essential IoT properties?
   a) Interconnectivity and artificial intelligence
   b) Heterogeneity, dynamism, and unlimited power supply
   c) Dynamism and artificial intelligence
   d) Interconnectivity, heterogeneity, and dynamism
3. The IoT focuses on
   a) Data visualization
   b) Data analysis
   c) Cross-platform design
   d) All of the above
4. Which of the following is not a stage in the IoT life cycle?
   a) Intelligence
   b) Knowledge
   c) Information
   d) Mobility
5. The challenges in building IoT systems are
   a) Technology and cross-platform solution
   b) Mathematical modeling of sensor data routing
   c) Designing machine learning-based applications
   d) None of the above

Conceptual Questions

1. How does sensor-cloud enable the IoT?
2. How are the life cycles of sensor-cloud and the IoT connected?
3. What are the major virtualization models in the IoT?
4. What are the major research challenges to fulfilling the properties of IoT?
5. Is the architecture of the sensor-cloud inclined to that of the IoT? If yes, then how?

References

[1] J. Morgan, "A Simple Explanation of 'The Internet of Things'," Online: www.forbes.com/sites/jacobmorgan/2014/05/13/simple-explanation-internet-things-that-anyone-can-understand/#3641450f1d09, 2014.
[2] K. Patel and S. Patel, "Internet of Things-IOT: Definition, Characteristics, Architecture, Enabling Technologies, Application & Future Challenges," *International Journal of Engineering Science and Computing*, Vol. 6, May 2016, doi: 10.4010/2016.1482.
[3] K. Karthikeyan and L. Sujatha, "Fluorometric Sensor for Mercury Ion Detection in a Fluidic MEMS Device," *IEEE Sensors Journal*, vol. 18, no. 13, pp. 5225–5231, July 2018, doi: 10.1109/JSEN.2018.2840331.
[4] L. Wang, A. Noureldin, U. Iqbal, and A. M. Osman, "A Reduced Inertial Sensor System Based on MEMS for Wellbore Continuous Surveying while Horizontal Drilling," *IEEE Sensors Journal*, vol. 18, no. 14, pp. 5662–5673, July 2018, doi: 10.1109/JSEN.2018.2840162.
[5] H. Yan, X. Liao, C. Chen, and C. Li, "High-Power Handling Analysis of a Capacitive MEMS Power Sensor at X–Band," *IEEE Sensors Journal*, vol. 18, no. 13, pp. 5272–5277, July 2018, doi: 10.1109/JSEN.2018.2839687.
[6] M. Gemelli, "Smart Sensors Fulfilling the Promise of the IoT," 2017. Online: www.sensorsmag.com/components/smart-sensors-fulfilling-promise-iot [Accessed October, 2017].
[7] S. Misra, S. Bera, and T. Ojha, "D2P: Distributed Dynamic Pricing Policy in Smart Grid for PHEVs Management," *IEEE Transactions on Parallel and Distributed Systems*, vol. 25, 2014, doi: 10.1109/TPDS.2014.2315195.
[8] Research and Markets, "Global Fingerprint Sensor Market 2018–2023: Proliferation of Fingerprint Sensors in Smartphones and Other Consumer Electronics," 2018. Online: https://tinyurl.com/ybc4yssf [Accessed May, 2018]
[9] P. Naiya, "Sensors in Smartphones to Top 10 Billion Unit Shipments in 2020," 2017. Online: www.counterpointresearch.com/sensors-in-smartphones-to-top-10-billion-unit-shipments-in-2020/ [Accessed December, 2017]
[10] J. Webb and D. Hume, "Campus IoT Collaboration and Governance Using the NIST Cybersecurity Framework," *Living in the Internet of Things: Cybersecurity of the IoT – 2018*, pp. 1–7, London 2018, doi: 10.1049/cp.2018.0025.
[11] J. Rajamaki, "Industry-University Collaboration on IoT Cyber Security Education: Academic Course: Resilience of Internet of Things and Cyber-Physical Systems," in *Proceedings of IEEE Global Engineering Education Conference (EDUCON)*, Tenerife, 2018, pp. 1969–1977, doi: 10.1109/EDUCON.2018.8363477.
[12] S. Benedetto, "The President's Page: IoT Initiative Poised for Growth in Collaboration with ComSoc," *IEEE Communications Magazine*, vol. 53, no. 2, pp. 6–9, Feb 2015.
[13] L. Guijarro, V. Pla, J. R. Vidal, and M. Naldi, "Game Theoretical Analysis of Service Provision for the Internet of Things Based on Sensor Virtualization," *IEEE Journal on Selected Areas in Communications*, vol. 35, no. 3, pp. 691–706, March 2017.
[14] S. Bose and N. Mukherjee, "SensIaas: A Sensor-Cloud Infrastructure with Sensor Virtualization," in *Proceedings of the 3rd IEEE International Conference on Cyber Security and Cloud Computing (CSCloud)*, Beijing, 2016, pp. 232–239.

[15] M. H. Shuvo and Y. Fu, "Sonar Sensor Virtualization for Object Detection and Localization," in *Proceedings of IEEE SoutheastCon*, Norfolk, VA, 2016, pp. 1–8.
[16] J. Kirklan, "Internet of Things: Insights from Red Hat," 2015. Online: https://developers.redhat.com/blog/2015/03/31/internet-of-things-insights-from-red-hat/ [Accessed March, 2015].
[17] A. Alamri, W. S. Ansari, M. M. Hassan, M. S. Hossain, A. Alelaiwi, and M. A. Hossain, "A Survey on Sensor-Cloud: Architecture, Applications, and Approaches," *International Journal of Distributed Sensor Networks*, Vol. 2013, 2013, doi: 10.1155/2013/917923.

FROM THE CORE TO THE EDGE: FOG

Chapter 8

Fog: The Next-Gen Cloud

In recent years, cloud computing has added a new dimension to the traditional means of compute, data storage, network management, and service provisioning. The advent of cloud computing [1–3] has been proven to be a major "boon" to the ever-growing cyber-physical world. Cloud services rely strongly on the centralized data centers (DCs) and the underlying IP-based network infrastructure. The key to the success of the cloud-computing paradigm are on-demand, real-time service provisioning and service-abstraction for the end-users and organizations. Cloud computing frees its end-users and user-organizations from complications and underlying architecture, which allows the users to enjoy the cloud-services seamlessly on an on-demand basis. While at one end, these salient features make cloud computing an intriguing and highly promising paradigm, it has also become the root of problems for a specific set of applications on the other end. It is understood that the use of smart, Internet-connected devices and increase in the demand for low-latency, real-time services pose a serious challenge to the traditional cloud-computing framework.

Rapid increases in the number of ubiquitous [4,5] and mobile sensing [6,7] devices connected to the Internet results in a steep rise in the volume of data to be processed, managed, and stored in the cloud DCs. Moreover, in light of emerging IoT solutions, many devices are located at the edge of the network, thereby requiring support for device mobility and demanding low latency, real-time, location-aware services [8]. Cloud computing falls short in multiple facets, when it comes to serving the ever-surging requests generated from billions of devices at the network edge, demanding latency-sensitive, location-aware services. Fog computing comes as a welcome solution to the aforementioned problems. Coupled with the cloud, this new and disruptive technological paradigm resolves *most* (if not *all*) of the challenges poised before it. In this chapter, we introduce the fog-computing paradigm and discuss its salient features.

8.1 Introduction to Fog Computing

Recent advancements in computer technologies have led to the conceptualization, development, and implementation of cloud computing systems [7]. Since its inception, the story of cloud computing has been mostly a successful one. Cloud-based solutions are highly reliant on data centric networks, which are treated as the monopolized hubs responsible for processing, computation, and storage. For all contemporary cloud-based solutions, service requests and resource demands are analyzed and processed within the DCs. A parallel growth in the number of end-user devices, the degree of heterogeneity of these devices, and the diversity of application requests pose a significant challenge to this centralized framework. As reported by Cisco, by the year 2020, the count for next generation Internet-connected (IoT-based) devices will rise to a whopping 50 billion [8]. In this reference, recent studies have identified fog computing as the technology for the IoT-centric cyber-future.

"FOG," the term was coined by Cisco employee Ginny Nichols [9] in mid-2012, stands as an abbreviation of "From Core to edge." The term *fog* was suggested as a metaphor – just like fog in nature is formed close to the ground, fog computing points to computation closer to the network's edge. The term also serves as a perfect foil to *cloud* computing, which takes place away from the network edge. Fog computing was primarily envisioned as a distributed

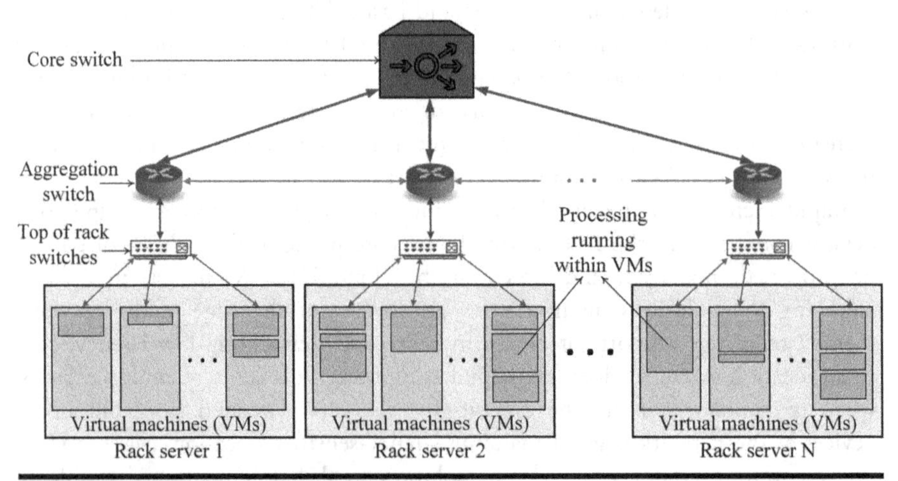

Figure 8.1 Skeleton of a cloud data center

computing framework, which empowers the network devices at different hierarchical levels with various degrees of computational and storage capabilities [10].

8.1.1 Where Does Cloud Computing Fall Short?

Before we plunge into the intricacies of fog computing, it is only fair to analyze exactly where cloud computing fell short and by what extent. In order to seek the answer to these questions, we must revisit the working principles of cloud computing. The fundamental characteristic that makes cloud computing such a promising technology is *virtualization* [11,12]. Figure 8.1 projects the internal structure of a cloud DC. A cloud DC is composed of multiple rack servers, each hosting a number of virtual machines (VMs). Note that a particular VM serves only one end-user by fulfilling all his/her requirements. Also, the number of processes running in the VMs may vary as may the resource (memory, processing cores, and network) requirements of each VM. All servers within a particular rack are connected by a top of rack switch. These switches, and in turn the rack servers, may communicate with one another over the aggregation switch, which is located in the layer immediately above the rack switches. The aggregation switches are further connected to a core switch, which acts as a gateway for inter-data center communication.

The primary unit responsible for virtualization in a cloud-computing framework is known as the hypervisor. A hypervisor, also commonly known as a virtual machine monitor [13–17], is responsible for the creation and running of the VMs and may exist as hardware, firmware, or software within the platform. Figure 8.1 shows the internal details of virtualization taking place within a cloud server. The VMs run on a guest OS (e.g., Linux, Windows, OS X, etc.), on top of which the application instances run. The VMs are connected through a virtual adapter to the virtual switch within the hypervisor. The hypervisor is also linked with the processing unit, which governs all operations within the hypervisor. The virtual switch within the hypervisor is connected to a physical adapter in the server, which links the server to the top of the rack switch. The key to virtualization is that it abstracts the users of the system to remain “blind” to the internal processing of their requests and yet allows them to enjoy the services seamlessly and at an on-demand, pay-per-usage basis. Therefore, in a cloud-computing framework, the distance between the user being served and the corresponding service DC can be significantly large. It is observed that with the increase in the number of concurrent service requests, the centralized computing architecture of the cloud struggles to meet latency requirements. In other words, with billions of Internet-connected devices simultaneously requesting real-time, latency-sensitive services from the cloud, classical cloud computing architecture fails deplorably to quench all their demands.

The fact that the cloud framework thrives on the idea of virtualization of services and relies on a highly centralized computing framework is the root cause of this major shortcoming of cloud computing. In the process of virtualization, the CSPs render their users with services from geo-spatially remote locations and thereby invoke high latency in service provisioning [8]. Also, as conventional cloud computing involves processing, computation, and storage of data only within DCs, the massive data traffic generated from the multitude of devices are anticipated to experience a huge network bottleneck and, in turn, high service latency and poor QoS [7]. It is often a waste of network bandwidth and time to transmit huge chunks of data from the numerous end-devices to the cloud-computing core and then transmit the responses from the cloud to the network's edge. Moreover, in order to process and serve this large number of requests, the DCs are required to be up and running around the clock, which results in the consumption of an enormous amount of energy and a huge carbon footprint. Figure 8.2 shows the role of hypervisor in providing support for virtualization within a cloud infrastructure.

Therefore, the need for a new computational paradigm is capable of handling the tsunami of latency-sensitive service-requests is indispensable. Herein, fog computing comes as an efficient solution to many of the problems the cloud had. Fog computing extends multiple functionalities of the classical cloud

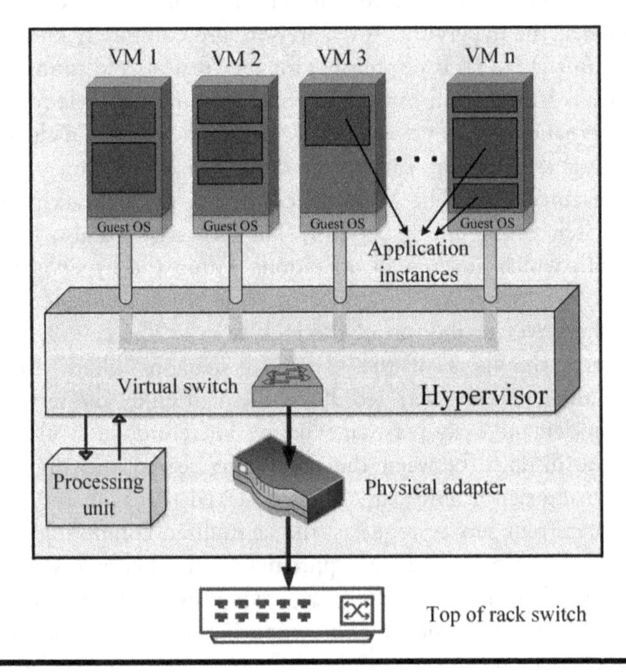

Figure 8.2 Virtualization within cloud

computing paradigm to the edge of the network – closer to the end-users and the mobile end-devices.

8.1.2 Definition

The pointers from the discussions on how to make cloud computing compatible with the upcoming IoT-based technologies all pointed towards one thing: decentralization. The goal was to relieve the cloud-computing core from being bombarded by the myriad of service requests at every instant by offloading the tasks to devices at the network's edge. Fog computing can be visualized as a non-trivial extension of the cloud-computing paradigm. Bonomi *et al.* [18] were early pioneers of research relating to fog computing, and their definition of the computing paradigm goes as follows.

> Fog Computing is a highly virtualized platform that provides compute, storage, and networking services between end devices and traditional cloud computing data centers, typically, but not exclusively located at the edge of network.

Although this definition may appear to be slightly ambiguous and structurally elementary, it highlights a couple of key characteristics of the fog-computing paradigm. First, the definition reveals that both cloud and fog use the same set of resources for computing, storage, and networking. Also, like the cloud, fog thrives on virtualization and shares many of the cloud's properties and attributes. Virtualization in a fog-computing scenario, however, is subtly different from that in the classical cloud-computing framework. Second, the definition states that the fog computing layer is an intermediate layer between the devices at the network's edge and the cloud computing core. A more comprehensive definition of fog computing is, however, proposed by Vaquero and Rodero-Merino [19], which is:

> Fog computing is a scenario where a huge number of heterogeneous (wireless and sometimes autonomous) ubiquitous and decentralized devices communicate and potentially cooperate among them and with the network to perform storage and processing tasks without the intervention of third-parties. These tasks can be for supporting basic network functions or new services and applications that run in a sandboxed environment. Users leasing part of their devices to host these services get incentives for doing so.

The definition asserts that fog computing is essentially a ubiquitous and decentralized computing paradigm capable of performing a significant number of tasks without the intervention of an established computing paradigm, such as the cloud. It also emphasizes on the aspect of device heterogeneity; this indicates that a fog-computing device can be located at the edge of the network (e.g., routers, switches, and access points). However, the definition does not highlight certain defining characteristics, such as service virtualization, geo-distributed nature, and large-scale latency-aware application service.

Considering the key attributes and properties of fog computing, we propose the following definition.

Fog computing is a decentralized, geo-distributed paradigm that relies on the principles of virtualization and service categorization to provide latency-sensitive and locality-specific services to a multitude of devices located at the edge of the network. The fog-computing layer is essentially constituted by multiple heterogeneous and autonomous network devices (such as routers, switches, and access points) located in an intermediate layer between the end-devices at the network's edge and the cloud data centers at the core of the network. Services offered by fog include computing, storage, and networking solutions.

The network devices in the fog computing layer are located at different hierarchical levels between the core and the edge of the network and are empowered with various degrees of computational, storage, and processing

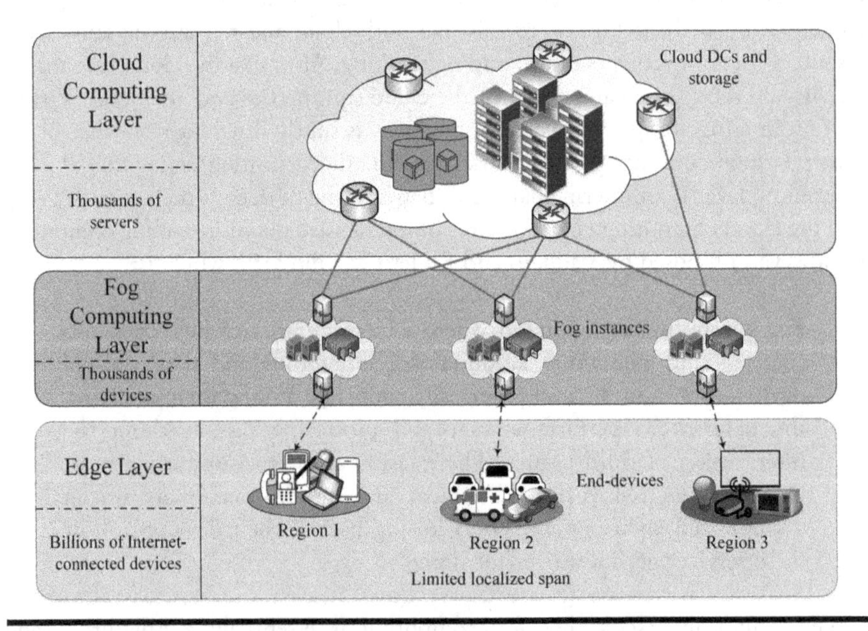

Figure 8.3 Generic outline of the fog-computing paradigm

capabilities. As stated in [7], these devices are equipped with intelligence, which allows them to examine whether an application request requires the intervention of the cloud computing tier. The idea is to serve the requests that demand real-time, low-latency services (e.g., live streaming, smart traffic monitoring, and smart parking) by the fog computing devices and connected work stations and small-scale storage units. For requests that demand semi-permanent and permanent storage or require extensive analysis involving historical data sets (e.g., social media data, photos, videos, medical history, and data backups), these devices act as routers or gateways to redirect the requests to the core cloud computing framework. Figure 8.3 [7,8] presents an outline of a generic fog computing platform. In the top layer is the core cloud computing module powered by high-end computing servers and data storage facilities. The intermediate layer comprises several fog instances, each of which is constituent of multiple fog computing devices [20,21]. The fog instances are connected to the cloud framework through one or more cloud gateways. It is mentionworthy that each fog instance only serves a particular geographical area or locality. In the lowest layer, the end-devices located at the edge of the network are shown. Note that the devices are clustered based on their locality, and every locality is served by a particular fog instance, which is dedicated for the purpose.

8.1.2.1 Is Fog a Replacement for Cloud?

As a continuation to this discussion, we address one of the most crucial (and popular) questions regarding fog computing: *Is the fog computing paradigm a replacement for cloud computing?* The one-word answer to this question is *no*! Fog computing is more like an add-on, an enabling technology for cloud computing. Judging by the different types of applications served by fog and cloud, it is clear that fog computing is not a replacement for cloud computing; these two technologies complement each other [7]. The complementary functions of cloud and fog enable end-users to experience a new breed of computing technology that serves the requirements of the real-time, low-latency applications running at the network edge and also supporting complex analysis and long-term storage of data at the core cloud-computing framework. More on the fog-cloud interplay from an architectural point of view is discussed in Section 8.1.4.

8.1.3 Fog versus Other Computing Paradigms

We now compare and contrast fog computing with other related computing paradigms. Apart from the classical edge computing and cloud-computing models, we analyze the correlation between fog computing and related mobile computing paradigms. To illustrate the differences between the computing paradigms, we take the example of a smart car parking system. We imagine a parking lot in which each parking spot is monitored by a parking-sensor checking on whether the slot is occupied. Each parking spot is also equipped with a signal-post that shows a green signal if the spot is vacant or a red one if it is occupied by a car. Change in the signal in a signal-post is triggered by the output of the corresponding parking-sensor.

8.1.3.1 Fog versus Edge Computing

Fog computing is confounded on the classical mode of edge computing. It is not rare that one fails to appreciate the novelty within fog computing, as the terms fog computing and edge computing are used interchangeably more often than not. In reality, while both of these technologies have many properties in common, and fog actually relies on the principle of edge computing itself, the difference between the two is subtle yet significant.

Edge computing relies highly on in-device, localized processing of service-requests (i.e., instead of having the processing done in the central server, edge computing draws computing to the end-devices, such as mobile phones and programmable automation controllers). This ensures a higher degree of operational independence and hence concedes fewer points of failure. Figure 8.4 shows the generic bilayer architecture of the edge-computing paradigm. It is observed that, similar to fog computing (Figure 8.3), edge devices are clustered by their spatial positions. Each such cluster is connected to a local edge-processing unit. The edge computing and storage units are distributed in the top layer of the architecture. However, due to the static clustering policy, in edge computing one computing cluster may have to serve devices of multiple regions. Also, it is pivotal that the architecture does not feature a centralized computing core, such as the cloud. In the absence of such a centralized, highly powerful computing platform, all computations are executed in the end-devices only. Requests that cannot be served by the edge devices are transmitted to the edge servers for further processing. The primary drawback of edge computing is that it requires the end-devices to perform continuous real-time categorization of the application requests and process those that can be served locally. This requires a constant supply of energy, which is often an adversity considering the limited power supply (mostly battery-driven) of the terminal nodes and the difficulty associated with regular replacement of the power-source. Also, due to the non-existence of a centralized core, it is extremely difficult and costly to obtain global data and trend analysis in an edge computing platform. Considering the example of the smart car parking system, in an edge computing platform, the parking sensors themselves assess the status of the parking spots (whether a car is occupying the spot or not) periodically and send messages directly to the associated signal-posts once there is a change in the status.

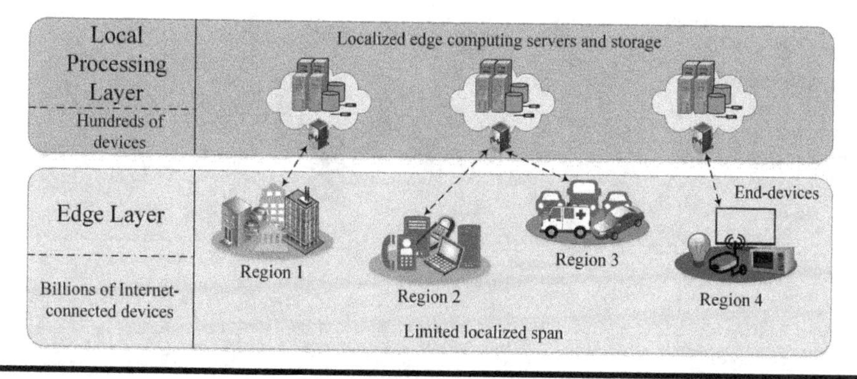

Figure 8.4 Generic outline of the edge computing paradigm

Fog computing, on the other hand, empowers the different network devices, such as routers and switches, with additional capabilities to compute and store. Data gathered by the end-devices are periodically sent to these devices, where the processing takes place. Fog frees the end-nodes from the overhead of performing the preliminary computation and subsequent decision making within the nodes themselves. Thus, the need for constant or frequent power supply to the terminal nodes is mitigated. All relating computation and processing take place within the fog devices, which already have an uninterrupted supply of energy. However, this architecture has a larger number of points of failure than its edge computing counterpart. In the context of the smart car parking system, in a fog computing solution, the parking-sensors simply sense and report the status of the parking spots periodically to their nearest router or network switch without doing any processing on the sensed data. The fog device, which in this case is a router or a network switch, performs the necessary analysis on the acquired data and eventually instructs the signal post, if there is a need to toggle.

8.1.3.2 Fog versus Cloud Computing

In simple terms, fog is a non-trivial extension of the cloud. Technologically and in principle, therefore, fog and the cloud are akin. Both of these technologies thrive on the principle of virtualization and allow their users to enjoy seamless service at an on-demand and pay-per-use basis. There are, however, certain dimensions in which these two computing paradigms vary – subtly, yet significantly. In Figure 8.5, the skeleton of the classical cloud computing paradigm is presented. While the top layer comprises cloud DCs and is responsible for the service of all application requests, the bottom layer consists of billions of devices requesting a variety of services. All computation, processing, analysis, and storage,

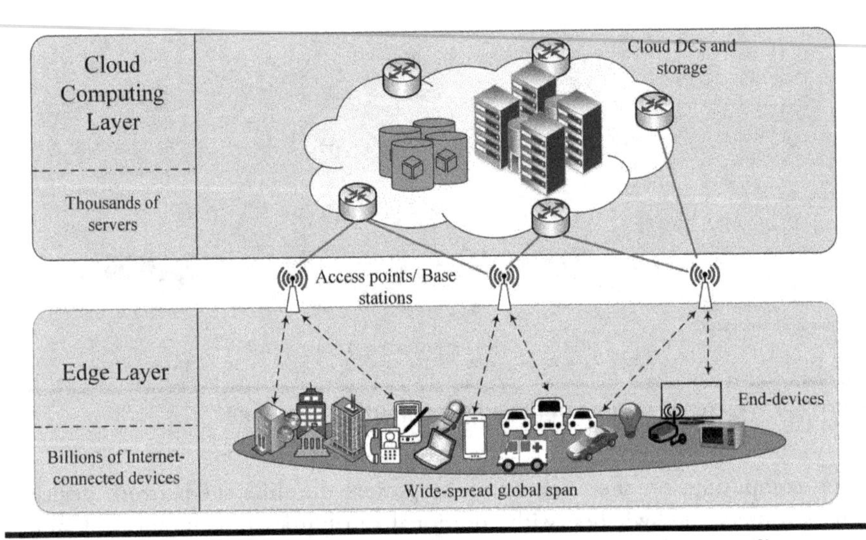

Figure 8.5 A generic outline of the traditional cloud computing paradigm

however, take place within the cloud-computing core. The key differences between fog and cloud are listed below.

- *Locus of computing*: The fundamental difference between the fog and cloud computing paradigms is the location of computation and storage. While a lion's share of the computation, storage, and management issues is taken care of at the centralized DCs in the cloud, fog draws out these functionalities close to the edge of the network. While, technically stating, computation does not take place at the network's edge (as in within the end-devices), it does take place in a highly decentralized fashion at an intermediate layer, This intermediate layer, often referred to as the fog computing layer, comprises a heterogeneous set of "intelligent" devices capable of performing necessary computations and processing data; it is empowered with small-scale temporary storage.
- *Aperture*: The centralized architecture of cloud computing allows it to have access to data across a larger span of area, and allows the cloud-computing core to perform global analysis of the acquired data. Having a global view of all the gathered data permits the cloud to achieve location-specific characterization and assessments. However, the amount of data to be migrated to serve such requests would cause tremendous overhead. However, as the fog computing layer is located close to the edge of the network, it serves as an ideal platform for location-specific analysis of data. In order to serve a location-specific request, the amount of data processed within the fog

computing layer is not overwhelmingly large. Fog, however, fails to provide a global overview of the system due to its myopic aperture.

- *Service latency*: In the context of serving locality-specific application requests, the service-latency in fog computing architecture is considerably higher than in classical cloud computing. Considering the amount of data which is to be migrated and managed to serve such a request, the service latency and service cost would be significantly high in the cloud. However, under similar constraints, fog is capable of providing near real-time service to latency-sensitive application requests, as the distance traversed by the data, is considerably lower, as is the amount of data processed.
- *Dimension*: Cloud DCs are usually very large, comprising thousands of servers. Each server is empowered with cutting-edge computing technology, data handing capabilities, and connectivity with large-scale storage units. On the other hand, fog computing devices are often perceived as different networking devices entrusted to perform a few additional tasks. These devices are capable of performing preliminary computing tasks and are supported by a limited amount of storage.
- *Composition*: While services in a cloud-computing framework are executed at the dedicated large-scale server machines, devices in a fog computing platform are more heterogeneous. A fog device can be anything, starting from a simple edge router or network switch to an access point or base station.
- *Network bandwidth*: It is imperative that devices requesting services from the cloud must be connected to the Internet. Clearly, the higher the bandwidth of the connection, the better the QoS. Services from the fog computing layer, however, are partially able to be extended without the end-devices getting connected to the Internet. Considering that a router or switch may act as a fog device, end-devices connected to it across local area network may obtain services from the fog layer without access to the Internet. The catch in this case is that services that require the intervention of the cloud-computing core, such as comparison with historical data and global trend analysis, cannot be offered without connectivity to the Internet.
- *Deployment*: Efficiency of the cloud-computing framework vastly depends on the arrangement of the cloud servers within the DCs. This calls for detailed and sophisticated planning. Deployment, organization, and connectivity among the fog devices are more ad-hoc in nature, which requires minimal or no planning.
- *Operators*: The establishment and maintenance of the cloud computing facility require a high degree of intervention of field-experts. Fog, on the other hand, requires minimal human intervention, especially in the maintenance phase.

Table 8.1 presents the differences between the cloud and fog.

The primary modes of communication and the location of computation for the three major computing paradigms are shown in Figure 8.6.

Table 8.1 Comparison of the characteristics of fog and cloud computing

Characteristics	*Cloud computing*	*Fog computing*
Locus of computing	Predominantly takes place within the centralized cloud DCs	A large proportion takes place within the fog layer, while the rest in the cloud core
Coverage	Wide-spread geographical coverage	Each fog instance monitors only a restricted area
Latency	The mean transmission latency and service latency are considerably high	Very low – especially for applications demanding real-time service with predictable latency
Dimension/ composition	Comprises of thousands of high-performance servers and data storage units	Relies on networking devices such as routers, switches, and access points, which are empowered with additional processing power and storage
Bandwidth	Demands high-bandwidth network connectivity among the cloud DCs	Internet connectivity is required to provide cloud-based services; otherwise capable of providing uninterrupted services with no or intermittent Internet connectivity
Deployment	Arrangement and connectivity among the servers are crucial for performance enhancement	Mostly ad-hoc in nature
Operators	Demands interventions of field experts	Mostly autonomous – requires minimal human intervention
QoS	Performance is compromised when the number of simultaneous service requests increases; often fails to meet low-end latency demands	Designed to provide better performance for latency-sensitive applications; improves overall system performance
Energy	Very high carbon footprint, as the cloud DCs usually run continuously around the clock	Highly energy-efficient; releases workload from the cloud, which allows the DCs to be idle
Cost	Very high deployment cost, coupled with high running cost and maintenance overhead	High establishment cost; however, incurs minimal cost overhead for execution and maintenance

8.1.3.3 Fog versus Mobile Computing

We now compare fog against the popular mobile computing paradigms (e.g., mobile edge computing and mobile cloud computing). The enabling characteristic of mobile computing is to provide support for mobile devices through

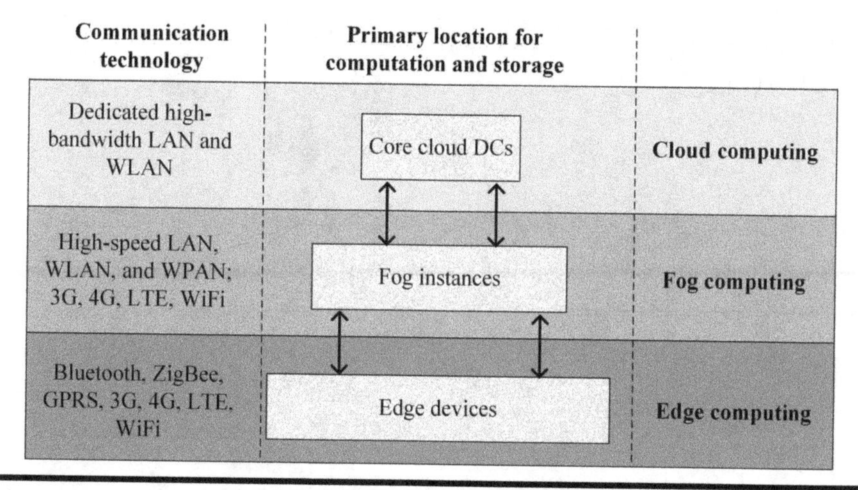

Figure 8.6 Comparison of different computing paradigms

ubiquitous and remote processing over the Internet. The storage primarily takes place within some servers, which are typically away from the end-devices.

For mobile cloud computing (MCC) [22–24] computations take place within the central cloud data centers (DCs). As shown in Figure 8.7, MCC architecture [25] is quite similar to that of fog computing. The difference, however, lies in the middle layer, which is as termed the mobile computing layer, in this case. Unlike fog computing, in MCC, each locality is served by a separate mobile network. Support of device mobility is ensured through handoff mechanisms.

On the other hand, mobile edge computing (MEC) [26,27] may be a reflection of MCC with servers located at the very edge of the network, instead of at the network's core. However, the computer hubs are much smaller than the cloud servers and deployed in higher numbers along the edge. A server in a MEC environment can be a mobile phone, a tablet, or a personal digital assistant. Note that MEC does not have a global computing framework (such as the cloud) [28,29] and therefore is capable of performing aggregated analytics on the data acquired across different geographic locations.

In contrast, fog computing can be regarded as a hybrid of mobile cloud computing (MCC) and mobile edge computing (MEC). Fog performs part of its processing closer to the network's edge, and the remainder is served by the core cloud computing module. However, from a technical perspective, it can be argued that fog is not a mere integration of these two technologies. The principal disruption about fog computing is that it adds a separate layer of computing and storage between the cloud layer and the network edge. Also, devices that perform the processing of the application requests at this intermediate layer are specially

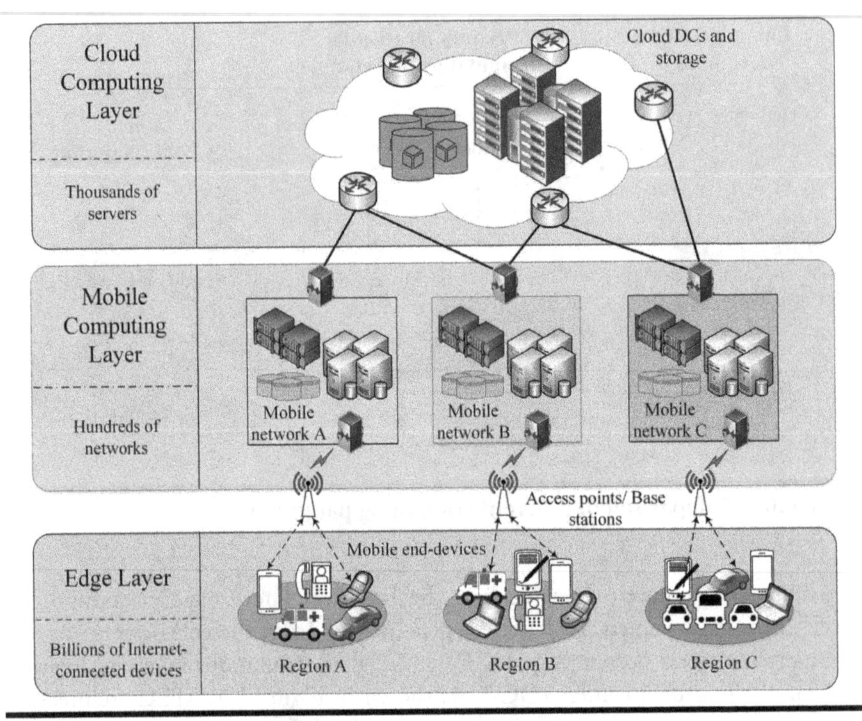

Figure 8.7 Mobile cloud computing (MCC) architecture

designed and manufactured. Mostly, the traditional network accessories, such as edge-routers and network switches, are reconfigured in a way to serve the purpose of screening off the incoming requests and serving the one demanding local, latency-sensitive processing.

8.1.4 Where There Is Fog, There Is Cloud

In this Section, we establish the juxtaposition of fog and cloud in the canvas of the future Internet. Fog computing is considered a natural, yet non-trivial, extension of the functionalities of the core of the network to its edge. While we have argued to establish the point that the concept of fog computing is truly disruptive and that this computing paradigm is significantly different from its close relatives, we must assert that fog computing is not a substitute for the classical cloud computing paradigm. Rather, fog and cloud complement each other in terms of service provisioning and data analytics. The very vision of fog computing was an amelioration of the classical cloud computing model

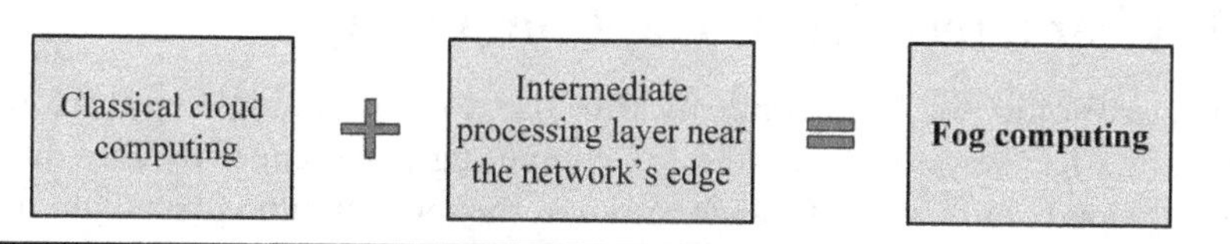

Figure 8.8 The inter-relation between cloud and fog

addressing the shortcomings of cloud [1]. Figure 8.8 gives a pictorial dimension to our discussion.

We highlight the inter-relation of fog and cloud computing below.

- *Cloud as part of fog*: Architecture-wise, fog acts as an add-on to the classical model of cloud computing, although the introduction of the fog layer takes away a significant task-load. All application requests that demand services involving long-term analysis of historical data or analysis of data gathered over a widespread area are redirected to the core cloud module. The cloud, therefore, remains the nucleus of the fog computing paradigm.
- *Inter-dependent work and data flow*: The flow of data and control within the fog and cloud modules overlap. Fog architecture demands the continuous exchange of control messages between the fog and the cloud computing planes. The flow of data, however, can be either unidirectional (from fog to cloud) or bidirectional, depending on the application type.
- *Minimizing points of failure*: As fog devices are equipped only with small-scale temporal storage, each device sends its aggregated data to the cloud DCs periodically. This policy is analogous to the memory write-back policy in case of operating systems. Although by bringing computing towards the edge, fog introduces multiple points of failure; as all data are regularly sent and replicated in the cloud DCs, the number of points of failure is reset on a regular basis. After each replication cycle, the count of possible points of failure in the system is minimized.
- *Mutual benefit*: In light of the aforementioned discussion, it is evident that both fog and cloud complement each other. Fog mitigates several shortcomings of the classical cloud computing model by providing location-aware services with predictable latency. On the other hand, the computing core within the cloud DCs allows fog to serve service requests requiring large-scale (either in terms of time or aperture) data analytics.

It is, therefore, appropriate to conclude that these two technologies conjointly give birth to a new breed of computing paradigm, which serves most of the requirements of the future Internet and enables efficient data management and processing.

8.2 Characteristics of Fog Computing

Characteristically, fog bears similarity with the cloud in multiple ways. However, there are a few ways in which the fog computing paradigm is notably different from its cloud counterpart. Characteristics of the fog computing paradigm follow:

(a) *Proximity to end-users*: In the fog computing framework, several services are hosted in close proximity to the end-users (i.e., at the edge of the network). This results in improved QoS and superior user experience.
(b) *Low service latency*: As the computing hubs are located in close proximity to the end-users, applications that require services with predictable latency are greatly benefited. In particular, applications demanding real-time services, such as live video streaming, smart traffic light monitoring, and ubiquitous healthcare, are served readily by the fog computing tier with no significant delay incurred.
(c) *Location-awareness service*: One of the primary features of fog computing is to provide location-aware services to the end devices. The fog layer is composed of several fog instances, each of which serves a small geographic region. With each fog instance processing only the requests generated from within a bounded area, services are provided with higher quality and within a predictable latency.
(d) *Dense distribution*: The lowest tier of the fog computing architecture, which comprises billions of Internet-connected devices, is highly distributed. It geographically spans a large area. Fog, therefore, is designed to handle the colossal volume of data generated from a large number of devices at high velocity and hence is compliant with the *big data* requirements.
(e) *Heterogeneity*: The fog devices, also sometimes referred to as fog nodes, may be of different varieties. In fact, they can be anything from a simple network switch to an edge router. Thus, the fog layer is considered highly heterogeneous in its composition. These devices, however, must be located close to the network's edge.
(f) *Support for device mobility*: Support for mobility of the Internet-connected devices is one of the key enabling features of the fog computing paradigm. While many of the end devices are mobile in nature, fog provides uninterrupted coverage for all these devices, irrespective of their location and mobility pattern. It also provides services across several verticals, such as information technology, entertainment, and healthcare, giving birth to a unique breed of service management for the next generation of Internet.
(g) *Dynamic scaling*: One basic demand of the future Internet applications is on-demand service provisioning. In other words, the application requests and in turn the end-users' demands are to be served instantaneously

without any significant delay incurred. In order to enable this feature, the fog architecture must be able to dynamically distribute its workload over multiple computing instances (also known as fog instances). These fog instances are configured in a way to split or merge based on the flow of the incoming service requests. A fog instance is capable of splitting into two, in case there is a heavy traffic load. Each of these split instances covers a smaller geographic area and provides increased QoS. When two or more adjacent fog instances are underloaded, they are merged together to form a single large computing instance. This way, all processing units, except one, may be turned off or put into the power-saving mode. The efficient and intelligent scaling mechanisms help in the improvement of CPU and network bandwidth utilization and reduction of the overall power consumption of the system to a significant extent.

(h) *Predominant wireless access*: Another key characteristic of the fog computing paradigm is wireless communication. Going by the modern trend, it is imperative that wireless sensors or more generically wireless communication plays a crucial role in the context of the IoT. The fog architecture is designed precisely to support wireless communications amongst the end devices and between the end devices and the network access points. Devices may operate in any standard wireless standards (e.g., Bluetooth (IEEE 802.15.1), Zigbee (IEEE 802.15.4), WiFi (IEEE 802.11), or other newer ones, such as IEEE 802.15.6 (for WBAN-based communication), and IEEE 802.15.7 (visible light communication)).

(i) *Interoperability*: Lastly, we focus on the operational aspects of fog computing. Interoperability among different providers is critical in case of fog. In order to provision seamless and uninterrupted support to the end-users, the fog computing architecture must be capable of tailoring and orchestrating the services acquired from several federated domains.

8.3 Advantages of Fog Computing

In this section, we analyze the technological edges which fog gives us over the traditional cloud computing model. In this section, we discuss the advantages of moving the computing load away from the core of the network and closer to the edge.

- *Improved QoS*: A glance at the block diagram of the fog computing architecture clearly shows that the fog computing layer acts like a semi-permeable filter that only allows certain application requests to reach the core cloud computing module. Applications with requests for permanent storage of data and large-scale data analytics are allowed a passageway to the cloud core, while a major chunk of application requests are handled by the fog layer itself. The fog computing layer, thereby, significantly diminishes the volume of network traffic. As a consequence, it also curbs the mean service latency by a considerable extent – especially for applications requiring real-time, localized processing. Fog computing, therefore, improves on the QoS ensured by the classical cloud computing framework in more than one way.

- *Cost optimization*: As a result of the reduced volume of data migration, the corresponding costs for transmission and migration of data are also reduced. Although it may seem that setting up the fog computing layer would incur an additional establishment cost, in reality, it is a short-term overhead. Effects of the one-time establishment cost fade out with time; in the long run, fog computing will prove to be highly cost effective compared to traditional cloud computing.
- *Restricting the bottleneck*: One of the biggest bottlenecks of cloud computing is its single point of failure. Loss of connection with the computing core or crashing of the cloud server poises severe threats in terms of loss and leakage of data. Although adding a new intermediate layer increases the number of points of failure, the framework becomes more resilient and tolerant to faults and failures. Fog computing, thereby, gives an advantage over the classical cloud computing model by restricting the network bottleneck.
- *Careful virtualization*: As we discussed in Section 8.1.1, cloud computing relies strongly on the principles of virtualization; eventually, it costs the cloud dearly in terms of service delay and service cost. Despite the fact that fog computing is also based on similar virtualization principles, processing and computations take place in a locality-specific manner. Fog computing effectively mitigates delay and cost overheads. As all application requests are first pre-processed in the intermediate fog layer, requests demanding latency-sensitive and locality-specific services are served immediately by the fog instances, thereby reducing the service latency and the overall service cost. Services provided by the fog layer are also virtualized. However, as the virtualization is highly geo-specific, it acts as an advantage in terms of service provisioning.

Having argued the benefits of fog computing over the classical cloud computing paradigm, it is only fair to state that fog computing performs best in the context of the IoT and the future Internet, where a massive number of devices are in need for locality-specific services in real time. In situations in which devices only look for historical data analysis or assessment of global trends or that require long-term storage of data, the classical cloud computing model is most suitable. The fog computing layer, in such scenarios, is an additional computational layer for screening the requests generated from the end devices before redirecting them towards the cloud DCs to be processed and stored. In a practical world, however, that is rarely the case. There is a mixture of two types of application requests. With each passing day, the proportion of applications requesting territory-specific, latency-aware services increases steadily.

8.4 Summary

In this chapter, we introduced the concept of fog computing and discussed its co-existence with the classical cloud computing paradigm. We also compared and contrasted fog computing with other related computing paradigms, such as edge computing and mobile computing, and highlighted their similarities and differences. The key characteristics and salient features of fog computing were also discussed. Finally, we listed the various advantages of fog computing over the traditional cloud computing model and established the suitability of the new computing paradigm. We concluded that, in the context of the modern and future cyber world, where device-to-device (D2D) and machine-to-machine (M2M) communication plays a crucial role, fog computing is a significant add-on to its cloud counterpart. However, cloud computing remains an integral part of the fog computing architecture. Rather than considering fog as a replacement of the cloud, it is contemplated that together these two disruptive technologies give birth to a new breed of computing facilities to serve the demands of the Internet-connected world.

Working Exercises

Multiple Choice Questions

1. Which of the following statements is not true regarding fog computing?
   a) Fog draws the cloud's ability towards the edge.
   b) Fog replaces cloud computing by all means.
   c) Fog provides real-time services.
   d) Fog can provide location-specific services.
2. Which of the following statements is false?
   a) Fog operates on a layer between the cloud and the edge.
   b) ISPs are potential fog service providers.
   c) Fog does not support device mobility.
   d) Fog reduces the computing load of the cloud.
3. Which one is not a characteristic of fog computing?
   a) close to user processing
   b) location-specific processing
   c) dynamic scaling
   d) none of the above

4. Which of the following is an advantage of fog computing over the cloud?
   a) cost reduction
   b) latency reduction
   c) none of the above
   d) both of the above
5. Which statement is false about fog computing?
   a) Fog computing is another name for edge computing.
   b) Fog and mobile cloud computing are different names of the same paradigm.
   c) none of the above
   d) both of the above

Conceptual Questions

1. How is fog a different computing paradigm from the cloud?
2. How does fog computing compare with mobile cloud computing and edge computing?
3. What are the main advantages offered by fog as compared to the classical cloud computing paradigm?
4. What are the defining characteristics of fog computing?
5. How do you think the IoT can benefit from fog computing?

References

[1] M. A. Vouk, "Cloud Computing: Issues, Research and Implementations," in *Proceedings of the 30th International Conference on Information Technology Interfaces*, Cavtat, Croatia, Jun. 2008, pp. 31–40.

[2] B. P. Rimal, E. Choi, and I. Lumb, "A Taxonomy and Survey of Cloud Computing Systems," in *Proceedings of the 5th International Joint Conference on INC, IMS and IDC*, Seoul, South Korea, Aug. 2009, pp. 44–51.

[3] S. Yu, C. Wang, K. Ren, and W. Lou, "Achieving Secure, Scalable, and Fine-Grained Data Access Control in Cloud Computing," in *Proceedings of IEEE International Conference on Computer Communications (INFOCOM)*, San Diego, CA, Mar. 2010, pp. 1–9.

[4] J. Hightower and G. Borriello, "Location Systems for Ubiquitous Computing," *IEEE Computer*, vol. 34, no. 8, pp. 57–66, Aug. 2001.

[5] D. Puccinelli and M. Haenggi, "Wireless Sensor Networks: Applications and Challenges of Ubiquitous Sensing," *IEEE Circuits and Systems Magazine*, vol. 5, no. 3, pp. 19–31, 2005.

[6] A. Pentland, "Looking at People: Sensing for Ubiquitous and Wearable Computing," *IEEE Transactions on Pattern Analysis and Machine Intelligence*, vol. 22, no. 1, pp. 107–119, 2000.

[7] T. Choudhury, S. Consolvo, B. Harrison, J. Hightower, A. LaMarca, L. LeGrand, A. Rahimi, A. Rea, G. Bordello, B. Hemingway, P. Klasnja, K. Koscher, J. A. Landay, J. Lester, D. Wyatt, and D. Haehnel, "The Mobile Sensing Platform: An Embedded Activity Recognition System," *IEEE Pervasive Computing*, vol. 7, no. 2, pp. 32–41, 2008.
[8] MarketWatch, "Cisco Delivers Vision of Fog Computing to Accelerate Value from Billions of Connected Devices," Online: www.theiet.org/resources/journals/research/index.cfm [Accessed Aug. 2014].
[9] N. N. Khan, "Fog Computing: A Better Solution for IoT," *International Journal of Engineering and Technical Research*, vol. 3, no. 2, pp. 298–300, 2014.
[10] J. Preden, J. Kaugerand, E. Suurjaak, S. Astapov, L. Motus, and R. Pahtma, "Data to Decision: Pushing Situational Information Needs to the Edge of the Network," in *Proceedings of IEEE International Inter-Disciplinary Conference on Cognitive Methods in Situation Awareness and Decision Support*, Orlando, FL, Mar. 2015, pp. 158–164.
[11] B. Sotomayor, R. S. Montero, I. M. Llorente, and I. Foster, "Virtual Infrastructure Management in Private and Hybrid Clouds," *IEEE Internet Computing*, vol. 13, no. 5, pp. 14–22, 2009.
[12] N. Liu, X. Li, and Q. Wang, "A Resource & Capability Virtualization Method for Cloud Manufacturing Systems," in *Proceedings of IEEE International Conference on Systems, Man, and Cybernetics*, Anchorage, AK, Oct. 2011, pp. 1003–1008.
[13] A. Desai, R. Oza, P. Sharma, and B. Patel, "Hypervisor: A Survey on Concepts and Taxonomy," *International Journal of Innovative Technology and Exploring Engineering*, vol. 2, no. 3, pp. 222–225, 2013.
[14] E. Bauman, G. Ayoade, and Z. Lin, "A Survey on Hypervisor-Based Monitoring: Approaches, Applications, and Evolutions," *ACM Computing Surveys*, vol. 48, no. 1, pp. 10: 1–10:33, 2015.
[15] Y. Suzuki, S. Kato, H. Yamada, and K. Kono, "GPUvm: GPU Virtualization at the Hypervisor," *IEEE Transactions on Computers*, vol. 65, no. 9, pp. 2752–2766, 2016.
[16] L. Cheng and F. C. M. Lau, "Offloading Interrupt Load Balancing from SMP Virtual Machines to the Hypervisor," *IEEE Transactions on Parallel and Distributed Systems*, vol. 27, no. 11, pp. 3298–3310, 2016.
[17] A. Blenk, A. Basta, M. Reisslein, and W. Kellerer, "Survey on Network Virtualization Hypervisors for Software Defined Networking," *IEEE Communications Surveys & Tutorials*, vol. 18, no. 1, pp. 655–685, 2016.
[18] F. Bonomi, R. Milito, J. Zhu, and S. Addepalli, "Fog Computing and Its Role in the Internet of Things," in *Proceedings of the 1st Edition of the MCC Workshop on Mobile Cloud Computing (ACM)*, Helsinki, Finland, Aug. 2012, pp. 13–16.
[19] L. M. Vaquero and V. Rodero-Merino, "Finding Your Way in the Fog: Towards a Comprehensive Definition of Fog Computing," *ACM SIGCOMM Computer Communication Review*, vol. 44, no. 5, pp. 27–32, 2014.
[20] S. Sarkar, S. Chatterjee, and S. Misra, "Assessment of the Suitability of Fog Computing in the Context of Internet of Things," *IEEE Transactions on Cloud Computing, in press*, 2016. doi:10.1109/TCC.2015.2485206
[21] S. Sarkar and S. Misra, "Theoretical Modeling of Fog Computing: A Green Computing Paradigm to Support IoT Applications," *IET Networks*, vol. 5, no. 2, pp. 23–29, 2016.

[22] H. Qi and A. Gani, "Research on Mobile Cloud Computing: Review, Trend and Perspectives," in *Proceedings of the 2nd International Conference on Digital Information and Communication Technology and it's Applications*, Bangkok, Thailand, May 2012, pp. 195–202.
[23] S. Misra, S. Das, M. Khatua, and M. S. Obaidat, "QoS-Guaranteed Bandwidth Shifting and Redistribution in Mobile Cloud Environment," *IEEE Transactions on Cloud Computing*, vol. 2, no. 2, pp. 181–193, 2014.
[24] M. Khatua, S. Misra, and M. S. Obaidat, "Quality-Assured Secured Load Sharing in Mobile Cloud Networking Environment," *IEEE Transactions on Cloud Computing, in press*, 2015. doi:10.1109/TCC.2015.2457416
[25] H. T. Dinh, C. Lee, D. Niyato, and P. Wang, "A Survey of Mobile Cloud Computing: Architecture, Applications, and Approaches," *Wireless Communications and Mobile Computing*, Wiley, vol. 13, no. 18, pp. 1587–1611, 2011.
[26] Y. Wang, M. Sheng, X. Wang, L. Wang, and J. Li, "Mobile-Edge Computing: Partial Computation Offloading Using Dynamic Voltage Scaling," *IEEE Transactions on Communications*, vol. 64, no. 10, pp. 4268–4282, 2016.
[27] S. Sardellitti, G. Scutari, and S. Barbarossa, "Joint Optimization of Radio and Computational Resources for Multicell Mobile-Edge Computing," *IEEE Transactions on Signal and Information Processing over Networks*, vol. 1, no. 2, pp. 89–103, 2015.
[28] S. Sarkar, S. Chatterjee, and S. Misra, "Evacuation and Emergency Management Using a Federated Cloud," *IEEE Cloud Computing*, vol. 1, no. 4, pp. 68–76, 2014.
[29] X. Chen, L. Jiao, W. Li, and X. Fu, "Efficient Multi-User Computation Offloading for Mobile-Edge Cloud Computing," *IEEE/ACM Transactions on Networking*, vol. 24, no. 5, pp. 2795–2808, 2016.

Chapter 9

Fog Computing Applications

Fog computing is conceptualized as a computation framework possessing the capability to serve a variety of applications. Fog computing is envisioned and designed as an infrastructure that extends the services of the centralized cloud computing framework to the edge of the network. By moving closer to the edge devices, fog computing is able to serve the incoming service requests in real time, without introducing any unnecessary delay. Fog computing, hence, is an attractive computational framework for serving latency-sensitive and real-time applications, which include emergency medical services, smart traffic management, smart transportation, real-time (live) content delivery, and online gaming and augmented reality (AR)-based applications. However, as the fog computing framework is invariably supported by a central cloud computing infrastructure, the services that require extensive and long-term data processing or global-scale data analysis are redirected from the fog computing tier towards the cloud data centers.

In this chapter, we present the different application domains and discuss in detail a few use cases that would benefit tremendously from the introduction of fog computing as their core enabling technology. First, we discuss the service model of fog computing in reference to the IoT, and then based on this model, we present the application domains of fog and show how fog computing could potentially benefit the applications.

9.1 IoT Applications and Fog Computing

Fog computing is often referred to as a key enabler of the IoT. The fundamental reason behind this statement lies in the design principles of the fog computing infrastructure. Unlike the classical cloud computing model, where computation and storage are supported by high-end computing units and storage disks, respectively, fog computing relies on devices that are typically constrained in terms of the (computational and storage) resources. These devices are typically envisaged as smart network devices, such as routers, switches, and gateway nodes [1–3]. Another realization of these fog computing nodes is the use of miniaturized computing nodes, such as the Raspberry Pi.

We show a typical fog-based service model in Figure 9.1. At the bottom-most layer of this service hierarchy, we have the edge device. This device is representative of any *thing* in the IoT ecosystem, ranging from "dumb" sensor nodes to

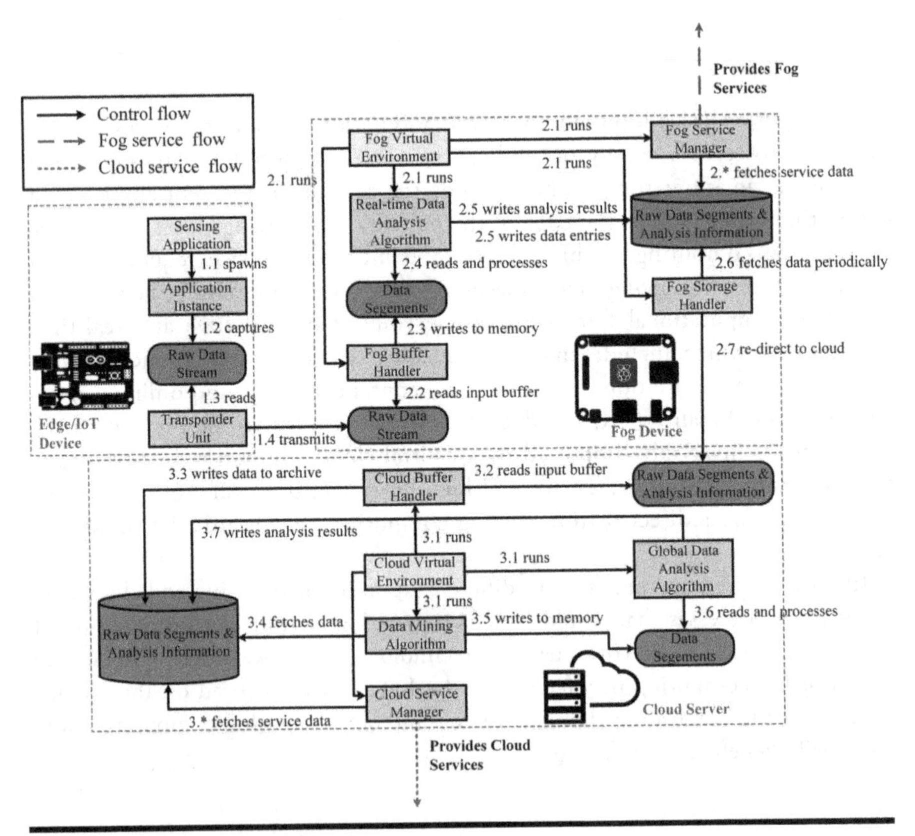

Figure 9.1 Fog service model

"smart" wireless healthcare devices. Regardless of their intelligence, all these devices have the two following characteristics: (a) they are capable of sensing some environmental or surrounding events and recording the data and (b) they are able to transmit the sensed data. In other words, these edge devices are usually equipped with a sensing module and a transponder module. A running instance of a sensing application installed on top of the edge device is responsible for the recording of the data sensed by the sensing module *(1.1)*, in Figure 9.1. In reference to the IoT ecosystem, data are considered to be generated (and subsequently processed) in the form of streams. The raw data stream recorded by the application instance is periodically written in the designated memory locations *(1.2)*. The transponder unit periodically reads the memory locations written by the sensing application instance and transmits it to the fog instance connected to it *(1.3)*. A fog instance is essentially a collection of several co-located fog computing devices.

Within a fog device (also known as a fog node), the fog virtual environment runs a number of applications, including the fog buffer handler, some real-time data analysis algorithm, the fog service manager, and the fog storage handler *(2.1)*. The fog buffer handler is responsible for reading the input buffer of the device, where the raw data stream transmitted by the edge device is received *(2.2)*. The buffer handler then segregates the input data stream into smaller meaningful data segments and writes the data segments in the main memory of the fog node *(3.2)*. The rationale behind the buffer handler writing the data segments to the main memory, and not to some secondary memory unit, is that these data segments are to be processed in real time and must be made readily available to the process performing the data analysis. After the data segments are written to the main memory blocks, the process/thread performing the real-time analysis of the input data reads the blocks and performs the necessary operations on the data segments *(2.4)*. This analysis includes data operations ranging from simple pre-processing or classification of data segments from extensive service-specific real-time data processing. Following this analysis, the processed data segments, along with the post-analysis derived data structures, are written back to the secondary fog storage unit *(2.5)*. A secondary fog storage is persistent, but is usually smaller in size as compared to that of a cloud data center. The small size of the fog secondary storage units mandates that the information is periodically moved to the cloud storage, based on the some retention algorithm (such as least recently used, least sparsely used, etc.) *(2.6, 2.7)*. As the fog tier primarily targets real-time service provisioning, there is very little benefit to keeping old data in the storage. On the other hand, periodic transfer of the locally processed information to the cloud enriches the quality of data in the cloud data center and thereby improves the quality of cloud analytics and service. The fog service manager, however, works in an ad-hoc manner (i.e., upon invocation by a service request generated by the end-user). Once a relevant service request is received, the fog service manager queries the storage unit, fetches the relevant service related information, and sends back the same in the form of the requested fog service *(2.*)*.

Within the cloud, the virtual environment runs the cloud buffer handler and cloud service manager along with the applications for data mining and analysis *(3.1)*. The raw data segments and analysis information re-directed to the cloud by the fog storage handler are received at the input buffer of the cloud. The cloud buffer handler is responsible for periodically checking the input buffer for new content. If some new content is encountered, the cloud buffer handler first reads the data from the input buffer *(3.2)* and then writes it back to the cloud storage unit *(3.3)*. At a later time, the data mining application accesses this storage, fetches necessary information *(3.4)* for processing, and loads it into the primary memory *(3.5)*. The global data analysis application then processes this information, based on the requirements of the application *(3.6)*, and subsequently writes the analysis result to the cloud storage *(3.7)*. The cloud storage units, thus, act as units of repository for the raw data segments, as well as the application-specific processed information and results. Similar to the fog service manager, the cloud service manager is designed to operate in an ad-hoc basis. Based on the service requests, it accesses the cloud storage units and serves the corresponding service requests *(3.*)*. Based on this representative service model for fog computing, we now put forward examples about how this service model can benefit the different application domains of fog.

9.2 Fog Applications

In recent times, fog computing has found its use in a wide spectrum of application domains, including advanced healthcare and well-being, smart cities, and multimedia. We discuss how the advantages of fog computing can be leveraged to seamlessly fulfill the specific service requirements of each of these application domains.

9.2.1 *Healthcare and Well-being*

Ubiquitous health monitoring [4] and e-health [5] are two of the most frequently used terms related to modern healthcare services. The underlying fundamental principle of both of these services is to enable automated and remote physiology monitoring. Often this is envisioned to be facilitated by the use of smart and wireless physiological sensors, also known as wireless body sensor nodes [6,7]. The sensor nodes are usually miniaturized, powered by batteries, and designed to monitor one or more physiological parameters, such as body temperature, heart rate, oxygen saturation level in blood (SpO_2), electroencephalogram and electromyogram signals, blood pressure, and blood sugar levels. These nodes are commonly worn as patches, as an on-body device supported by a band, belt, or vest. The devices also have, by design, a wireless communication module, such as Bluetooth or ZigBee, which allows them to periodically transmit the sensed data to a local data processing unit (LDPU), over wireless media. Recent developments include development of wireless communication standard (e.g., the IEEE 802.15.6 standard) [8,9], which is particularly designed to support ultra low-power, short-range, and efficient transmission of the medical data in a wireless body area network (WBAN). A WBAN is essentially a communication network encompassing a number of body sensor nodes; it operates in a coherent, cooperative manner. The data transmitted by these sensor nodes are gathered by the LDPU and are redirected to the computing core for faster processing.

Following the acquisition of the data at the computing core, two predominant service models are most commonly referenced:

- *Latency-aware service model*: The first model concerns fundamental processing of the input data and their representation in formats understandable by doctors and/or medical personnel. The data are processed in real time and presented to doctors, in remote hospitals, who analyze the data; necessary measures are taken thereafter [6,10]. This service model is relevant for medical services that require instantaneous actions assisted by the intervention of the doctors or medical personnel. These services include ambulatory health monitoring of patients [11], ubiquitous monitoring of patients suffering from chronic diseases [6], remote fall detection and health monitoring of elderly people [12], and hospital-based patient monitoring [6,11].
- *Latency-agnostic service model*: The second service model is automated and mostly unmanned; the physiological data are fused with the patient's previous data entries. Comparison of the recent data with some standard data set for anomaly detection can also be done. Analysis reports and relevant service results are sent back to the end-users (the patients, in this case). As this service model has high reliance on the data analysis algorithm and involves little or no human intervention, it is designed for services that are not critical or life threatening. Examples of services that can be handled by the service model include well-being services, such as smart-scale related services [13], fitness trackers, and activity monitoring [14]. These services typically do not require instantaneous processing of the data, and an algorithmic error or system failure is not expected to result in a medical catastrophe.

Traditionally, cloud computing has been considered to support the computing and storage services required to facilitate the processing of the data. However, the far-from-the-edge architecture mandates that the data must traverse all the way from its point of origin (at the edge of the network) to the cloud (at the core of the network) before it is processed, and then service data again has to travel back from the cloud core to the network's edge. This increases the overall service latency quite a bit. Moreover, with increasing numbers of services being reliant on the cloud for faster processing, the turnaround time for the tasks at the cloud is on the rise. Therefore, use of cloud as the computing infrastructure, particularly for services that require real-time processing of critical medical data, is not advisable [6,11]. Fog computing, in this context, is an appropriate platform; together with the cloud at the core, it can meet the service-specific requirements of the diverse spectrum of healthcare and well-being applications. Applications that follow the latency-aware service model (i.e., that require real-time data processing and latency-sensitive services) are served by the fog layer situated closer to the network's edge; those of the latency-agnostic service model are redirected to the cloud tier for necessary processing.

Based on these principles, a number of service models have been proposed in recent times. Masip-Bruin *et al.* proposed a service model for fog-based healthcare called Fog-to-Cloud (F2C) [5]. F2C is conceptualized to have multiple computing tiers between the fog tier, at the bottom of the hierarchy, and the cloud tier, at its top. These intermediate tiers, known as dynamic clouds, enable context-aware and patient-specific decision making with optimal resource utilization. Each is designed carefully to provide a specific set of services. In this work, through F2C, the authors provide five different services, which are (i) real-time oxygen-dose monitoring, (ii) real-time estimation of patient's efforts, (iii) therapy and treatment based on patient's activities, (iv) acquisition and subsequent processing of other relevant contextual information, and (v) context-sensitive therapy and treatment for patients. However, the number of such services, and correspondingly the number of tiers in the architecture, varies based on one's requirements. Another similar resource-efficient fog computing model for medical cyber-physical systems has been presented in [16], in which Gu *et al.* suggested a linear programming-based two-phase heuristic algorithm for optimal resource utilization and cost minimization.

A smart tele-healthcare framework, known as smart fog, was proposed by Borthakur *et al.* [17], where Raspberry Pi units are used to realize the fog computing tier. While the fog tier is responsible for performing machine learning-based smart decision making in real time, the cloud core is used for long-term data storage and global-scale data analysis. Craciunescu *et al.* [18] designed and developed an Arduino Uno-based fog computing framework to facilitate real-time patient health monitoring. The system is designed to sense and report the event of anomalies in heart rate and SpO_2, falling of patients, and even leakage of harmful gas. By offloading the processing load from the cloud tier to the fog tier, the authors have achieved significant improvement in terms of optimization of the service latency. The use of fog computing-driven service models for enhancement of security in data transmission in healthcare systems is also not out of place [19,20].

9.2.2 Smart Vehicle Management

Fog computing has enormous applicability in smart vehicle management systems. In a smart world driven by the IoT and emerging technologies, such as machine-to-machine (M2M) [21] and device-to-device (D2D) [22] communications and software-defined networking (SDN) [23], the concept of a vehicular network is gaining increased popularity each passing day. One key component of smart vehicle management is the vehicular ad-hoc network (VANET), fundamentally a network of connected vehicles. The nodes in VANET may communication among themselves over the Internet and even make decisions [24]. One popular realization of VANET is self-driving car technologies, such as Google's Waymo and Tesla's Autopilot and Summon [25], which are coming to reality. Another principal component of the smart vehicle management system is efficient traffic monitoring with the smart traffic light system [1,26]. We now discuss how fog computing can facilitate the realization of these component technologies and push towards a smart transportation system.

- *Vehicular ad-hoc network*: One of the key features of fog computing is support for mobility. This, coupled with the ability of real-time data processing and service provisioning, makes fog computing a suitable computing framework for supporting VANETs with a distributed set of sensors for monitoring parking spaces, weather conditions, and even road and traffic conditions. Hou *et al.* [27] proposed a vehicular fog computing service model in which they envision a cooperative environment of computing and communication resources available in vehicles and other edge devices, acting as the fog computing nodes. In the proposed model, the mobile vehicles constituting the fog tier and the base stations and the road-side units (RSUs) perform cloud operations. While this model may have some potential, major limitations could arise from the resource-constrained nature of the computing tiers. Neither the fog tier, formed principally by the vehicular resources, nor the cloud tier, assisted by the resources available at the base stations and RSUs, is expected to procure ample computing and storage resources to support the increasing demand.

  As an alternative solution, Truong *et al.* [28] proposed an SDN-supported VANET to leverage the benefits of fog computing and enable seamless vehicle-to-vehicle communications. In the fog-based SDN structure (FSDN) proposed in the work, the authors envision the mobile vehicles acting as edge devices.

The vehicles are connected to the regional fog instances by the cellular network and the fog instances constantly supported by an SDN controller. At the core of the network remains the cloud, which is also connected to the SDN controller. This model does not put much computational load of the vehicular resources, and with the SDN acting as an orchestrator between the fog and cloud tiers service classification, data processing and integration operations can be performed seamlessly and very quickly.

- *Smart traffic light system*: Improving the traffic management system is an important component of a smart transportation system, and smart traffic light control plays a crucial role. Bonomi *et al.* [1] presented an outline of a smart traffic light system (STLS), the primary objective of which is to maintain a steady flow of traffic along the main roads and prevent road accidents. At the bottom-most tier of the system model, a pre-deployed set of sensors is envisaged as the edge devices along with the mobile vehicles. The sensors operating as the edge devices are responsible for two primary phenomena – measuring the distance and speed of vehicles approaching the sensor from every possible direction and detecting pedestrians, cyclists, and individuals in wheelchairs near the roadside footpaths and crossing the roads. The operating principle is to determine in real time the optimal signal-change cycle based on the traffic flow pattern (i.e., the traffic density, the on-road vehicles' speed approaching from different directions, proximity of the vehicles, and the positions of the pedestrians and other slow moving entities). Clearly, as the system deals with mobile vehicles and involves human lives, the decision making is unambiguous, precise, and instantaneous. The location of the decision maker (DM) module, hence, must be closer to the edge in order to minimize the decision-making time. Alternatively, the DM can be conceptualized in a hierarchical manner, where a few modules are kept closer to the edge as part of the fog tier, while the others can be moved to the cloud. The four key attributes of the STLS, as envisaged by the authors, are: (i) support for wide geo-distribution of mobile devices (vehicles), (ii) predictable and low-latency services, (iii) consistent and unambiguous decision making, and (iv) multi-agency orchestration.

  Another secured traffic light control scheme is proposed in [26], in which the authors propose a location-based encryption scheme for secured message passing and truthful information sharing. While in VANETs, the vehicles are always supposed to honestly report their position and other relevant attributes, such autonomy can also act as a gateway for attackers, who can exploit this information and cause serious damage. It is, therefore, important to make sure that the information sent by the vehicles is truthful. In this context, the authors proposed a location-specific encryption scheme, which is governed by the smart signals in the locality. Once realized, these types of encryption schemes can be useful in securing the STLS.

9.2.3 Smart City Applications

The concept of the smart city has received a major technological boost in recent times with the advent of the IoT. With technology enabling inter-device communication (D2D and M2M), the idea of a smart city as a collection of individually smart components has a platform for its realization. The term *smart city* refers to a region that is technologically sustainable and comprises several independent smart modules designed to operate in a coherent and cooperative manner by communicating among themselves. Clearly, there are a lot of data processing operations and analysis and decision-making processes involved in each of these constituent smart modules. In an IoT ecosystem, data are generated in the form of streams, which results in the generation of a huge volume of data at a fast rate. Smart city applications require these data streams to be processed quickly, which are immediately translated into services results, in order to increase the QoS. Moreover, the localized nature of data processing makes transport of this huge amount of data to the cloud computing core for processing and storage an additional migration overhead. Fog computing, in such a scenario, fits the bill aptly, as it facilitates real-time processing of the incoming data streams locally, thereby providing the services quickly. Localized data processing implies that each fog instance has to process only a fraction of the total data volume; with data having to travel a smaller distance for processing purposes, fog computing provides the ideal platform to host and service smart city applications.

We discuss a few popular smart city applications, concentrating on smart technologies in the urban city context where some of the necessary infrastructure is already available.

- *Smart buildings*: Fog computing is the most suitable computational framework for meeting the service requirements of smart buildings. The term *smart buildings* includes smart homes, smart office spaces, and even smart hospitals. Dutta and Roy [29] presented a prototype of a smart building as a use case of the smart city application. Their idea was to associate different types of sensing units with different appliances in a home and facilitate inter-device communication by installing communication modules, such as Bluetooth or WiFi, within them. Thus, the edge tier comprises the usual home appliances, such as refrigerator, air conditioner, air purifier, lights, fans, and smoke detector, all of which are equipped with sensors and communication modules. With this setup, the appliances can now sense the requirement parameters and even communicate among themselves over Bluetooth or WiFi. The fog tier is envisioned to be located within the smart building as typically the routers and gateways are configured to act as the fog nodes. Given the sensitivity of the data, which are to be processed in a home-IoT environment, localized and building or home-specific deployment of fog computing nodes makes sense in terms of data privacy protection and data security. Any building-specific decisions are made in real time within this fog tier, and authorized users are able to have a visual representation of every ongoing activity in the building through smartphones. Using the smartphone, users can even control the operations of smart devices in their homes. The cloud tier can be reached over the Internet, and periodically the encrypted data are moved into the cloud for storage and global analysis purposes.

A similar service model has been proposed by Amadeo *et al.* [30]. The home appliances constitute the edge tier with a fog node specifically deployed for each home. However, instead of configuring a router or gateway device to act as a fog node, they suggested the use of Raspberry Pi units to constitute the fog tier. The use of Raspberry Pi in the fog tier is justifiable, as it has enough computing and storage ability to process the incoming data streams and quickly serve the application requests. However, configuring a router or a gateway to act as a fog device can be expensive when it comes to home-level deployment of the fog nodes. Similar models of fog computing to support smart home applications are suggested by other researcher groups as well [31,32].

- *Smart grid*: Application of fog computing to design and develop smart grid systems has been taken up with interest in recent times. Smart energy management and distribution are key aspects of smart cities, and the smart grid has made significant contributions. Okay and Ozdemir proposed a fog computing-based service model for smart grids [33,34]. In [33], they proposed a hierarchical communication structure in which the edge tier is constituted by smart meters installed within homes. The authors envisage four different types of intra-tier and inter-tier communications: (a) intra-edge tier communication among the smart meters in different homes, (b) inter-tier communication between the smart meters at the edge tier and the fog instances at the upper tier, (c) intra-fog tier communication among the different geo-spatially distributed fog instances with the fog tier, and (d) inter-tier communication between fog instances and the cloud computing core. As a smart grid allows two-way flow of electricity and information, intra-tier and inter-tier communications play an important role in its service model, and the smart meter plays the important role of monitoring these flows. However, in this process of flow monitoring and tracking, each smart meter produces huge amount of data continuously over time. The fog tier is responsible as a computing platform that provides support for collection, computing, and storage of this continuous data stream in real time. This not only facilitates efficient energy management and distribution, it also allows precise computation of the energy cost by taking into account real-time variations of the cost parameters.

As a continuation of this work, the authors later proposed a secure data aggregation scheme for fog computing-based smart grids, in the context of the IoT [34]. The scheme reduces the processing and storage load from the cloud, moving the tasks to the distributed fog instances and thereby achieving reduced service latency and computational overhead. The hierarchical model additive homomorphic encryption schemes could be applied to personal data, which may enhance the data privacy protection means and protect data from third parties. Another energy-aware and privacy-aware aggregation model was proposed by Lyu *et al.* [35], in which the fog instances periodically collect data from the connected smart meters and precisely derive fine-grained energy-flow statistics through aggregation within the fog tier.

- *Smart agriculture*: Connected sensor-equipped devices and the IoT play an important role in the areas of smart and precision agriculture; fog computing is an appropriate computing platform to support the IoT infrastructure. Ferrandez-Pastor *et al.* [36] proposed a precision agriculture method that is supported by a distributed computing architecture. First, a number of sensor nodes, embedded devices, and mobile devices are installed and configured to operate within the agricultural fields. This network of devices, along with the controlling subsystem, constitutes the edge tier of the service model. The data collected by the edge devices are then processed at the fog tier, where several agricultural data processing applications run and perform a wide range of processes, including irrigation control, sensor data management, image processing, and climate control. Based on the outputs of these processes, decisions and actions, prediction, planning, inference, and diagnosis follow. The entire process described is specific to an agricultural field, and hence, localized data processing within the fog tier makes sense, in improved QoS and data privacy protection. Munir *et al.* proposed a fog and cloud-assisted agricultural paradigm for the future IoT [37]. A layered fog computing architecture was suggested in this work, as the internal design of the fog tier. The fog tier was envisioned as a hierarchical combination of five layers: the hardware layer, the re-configuration layer, the virtualization layer, the analytics layer, and the application layer. Both models thrive on the edge tier at the bottom, comprising IoT devices, which seamlessly and continuously stream data from the agricultural fields, allowing for dynamic and real-time decisions to be made in the fog tier.
- *Smart waste management*: Efficient waste management and disposal is a key attribute of a modern smart city. With the garbage cans and dumps fit with

sensors, waste can be monitored in real time; once cans and dumps are sensed to be full or somewhat close to full, a garbage truck can vacate them during its next iteration. While the sensors in the garbage cans and dumps form the edge tier in this case, the fog tier is conceptualized as highly localized, spanning the routes of a few garbage trucks only. The data can be sent to the fog tier through modes of opportunistic communication or even crowd sourcing. Based on the incoming data, decisions are made and sent to the garbage collection trucks, which collect garbage only from the full cans. This optimizes the routes of the garbage trucks; more importantly, this ad-hoc mode of garbage collection ensures that garbage cans are never overflowing.

Similarly, fog computing preserves its suitability as a computing framework for other smart city applications, such as smart wind farms [1], fire detection and fire fighting, smart water management, and even greenhouse gas control.

9.2.4 Smart Data Management

Modern information and communication technologies (ICT) are continuously pushing towards achieving improved QoS, better user satisfaction, and efficient means of content management through cost-effective and sustainable solutions. A decade or so ago, cloud computing brought forward a revolutionary change by offering the concept of virtualization. Since then, cloud computing has always been an integral part of ICT and was the core component of a new breed of service models based on virtualization. However, with the emergence of the IoT, data management has posed a series of questions of the cloud framework. Data are now are generated in the form of streams at a high velocity. The data volume multiplies quickly, as does the amount of data generated. Hence, efficient, faster processing of data streams is becoming an increasingly challenging task for the cloud. The main factor is the increased distance of the cloud from the origins of data coupled with the heavy workload on the cloud. Fog computing, on the other hand, comes with a design that is extremely layered, allowing data to be processed very close to their origin. Moreover, by design, fog is more distributed than the cloud, and this geo-distribution of resources reduces the workload of the fog computing infrastructures. Fog computing, thus, has all the required characteristics to provide real-time services with improved QoS. We describe some of the principal application areas of fog computing in reference to smart data management.

- *Content delivery*: Content delivery and content caching are two key advantages of fog in terms of efficient content distribution. Lai *et al.* [38] proposed a QoS-aware streaming service assisted by fog computing infrastructures. Due to the limitations of cloud computing in providing time critical services, such as multimedia streaming and emergency notification, the fog computing platform was chosen for efficient content delivery. New content is periodically pulled by the fog-based content mirrors from the cloud and made available for edge devices. Therefore, when there is a new request for the latest content, it is not redirected to the cloud, where the most recent copy of the content is present; rather, the queries are served directly from within the fog tiers. An alternative way of implementing this service model is that whenever there is some new content available at the cloud, it is pushed down to the fog, and the fog mirrors are updated. Either way, the distance traversed by the service query, as well as the service latency, is reduced significantly. The number of copies fetched to serve incoming service requests is also significantly reduced once the content is cached in the fog nodes. A secured content-caching scheme for disaster management, based on this model, has been proposed by Su *et al.* [39]. Even though the cached contents are lost in an emergency, they can be refetched using the principle of redundancy.

- *Radio access network*: Radio access points can play a crucial role in the realization of fog computing and the IoT as technologies. The fact that the radio access points can be used as fog computing nodes makes them an interesting aspect of the fog computing paradigm. Fog radio access networks (F-RANs) are envisioned to enhance the performance of the cloud radio access networks through the enhancement of the remote radio heads in terms of processing and storage capabilities. F-RAN has the ability to pre-fetch the most recent copies of the most frequently requested files and cache them at the edge devices. Park *et al.* [40] proposed an optimized content delivery technique in a fog computing environment by modifying the content fetching and delivery schemes. A similar scheme has been suggested in [41]; these authors considered that the most frequented or the most sought content list might vary with time. Hence, they proposed an online caching scheme, assisted by the fog tier, based on the dynamic content popularity.
- *Big data analysis*: It is understandable that due to the benefits of localized and faster data processing ability, fog computing stands out for applications that involve big data processing. The intelligence of the fog computing nodes, supported by fog storage, makes fog the ideal platform for processing incoming big data streams in real time and only sending the aggregated data to the cloud. Particularly, with the geo-distribution of service points and support for device mobility, fog computing is far more compatible for supporting mobile big data applications than mobile cloud-based computing frameworks. It reduces the amount of data to be migrated and the cost of operation. Tang *et al.* [42] analyzed the impact of fog computing for big data processing with a smart city use case. They concluded that the multi-

layer service model of fog is able to provide the service more quickly at neighborhood, community, and city scales. Once the intelligence of the fog devices was enhanced further, performance was observed to improve even more. Sarkar *et al.* [1] reached a similar conclusion in their assessment of the suitability of fog computing to serve big data applications in the context of the IoT. They remarked that, as the percentage of services requiring real-time or latency-sensitive services increases in the pool, the efficiency of fog computing also increases.

9.2.5 Other Emerging Application Sectors

Fog computing has also found its applicability in several emerging technologies, such as online gaming [43], augmented reality-based applications [44], and real-time video processing-based applications [45]. Fog platforms are used to achieve higher quality of experience (QoE) in online gaming. Particularly for massively multiplayer online games, fog computing serves as the ideal platform, with its faster data processing, better content caching policies, and better content delivery principles. Lin and Shen recently developed a lightweight client, known as *CloudFog*, which extends the cloud's gaming engines to the fog and offers quicker video rendering and improved QoE for the players [43]. For similar reasons, fog computing is more often used to test augmented reality-based applications. Applications developed for Google Glass and Cloudlet are noted to be better experienced when supported by a local fog than some remote cloud infrastructure.

9.3 Summary

In this chapter, we introduced the service model of IoT applications in the fog computing paradigm. We, then, discussed the different fog applications in separate IoT domains (i.e., smart healthcare, smart vehicle management, various smart city applications, smart data management, and other emergency application sectors). For each application, we presented the service model and highlighted the role of fog computing in the specific context. We observed that due to its characteristic features, fog computing finds suitability in many applications that demand latency-sensitive and/or locality-specific data processing and service provisioning.

Working Exercises

Multiple Choice Questions

1. Which device is not a typical edge device?
    a) a high-end rack server
    b) a smartphone

   c) a temperature sensor node
   d) none of the above
2. Which device is not usually considered a typical fog node?
   a) a network router
   b) a Raspberry Pi
   c) a 3-layer network switch
   d) all of the above
3. Which of the following statements is false concerning the IEEE 802.15.6 communication standard?
   a) It is designed to support medical traffic communication.
   b) It has a wide communication radius.
   c) It can support high data rates up to 10 MBps.
   d) None of the above
4. What is not a part of a fog-assisted STLS?
   a) road-side units
   b) sensor-equipped vehicles
   c) wind mills
   d) a decision maker (DM) module
5. Which of the following statements is false?
   a) Smart meters are an essential part of smart grids.
   b) Fog computing can be used to improve the performance of cloud-assisted RANs.
   c) Both of the above
   d) None of the above

Conceptual Questions

1. How do you think fog computing can help in building a smarter city? Illustrate with three use cases.
2. What is the potential of fog computing for live streaming and content delivery applications?
3. Describe the fog service model and highlight the data lifecycle.
4. What do you understand by latency-aware and latency-agnostic fog service models in the context of fog-assisted healthcare?
5. What are some of the emerging application areas for fog computing, and why do you think fog computing can potentially help in enhancing the QoS for them?

References

[1] F. Bonomi, R. A. Milito, P. Natarajan, and J. Zhu, "Fog Computing: A Platform for Internet of Things and Analytics," *Big Data and Internet of Things*, pp. 169–186, 2014. Online: https://link.springer.com/chapter/10.1007/978-3-319-05029-4_7.

[2] S. Sarkar and S. Misra, "Theoretical Modeling of Fog Computing: A Green Computing Paradigm to Support IoT Applications," *IET Networks*, vol. 5, no. 2, pp. 23–29, 2016.
[3] S. Sarkar, S. Chatterjee, S. Misra, and R. Kudupudi, "Privacy-Aware Blind Cloud Framework for Advanced Healthcare," *IEEE Communications Letters*, vol. 21, no. 11, pp. 2492–2495, 2017.
[4] P. Hwang, C. Chou, W. Fang, and C. Hwang, "Smart Shoes Design with Embedded Monitoring Electronics System for Healthcare and Fitness Applications," in *Proceedings of IEEE International Conference on Consumer Electronics*, Nantou, Taiwan, 2016, pp. 1–2.
[5] A. Abbas and S. U. Khan, "A Review on the State-of-The-Art Privacy-Preserving Approaches in the e-Health Clouds," *IEEE Journal of Biomedical and Health Informatics*, vol. 18, no. 4, pp. 1431–1441, 2014.
[6] S. Misra and S. Sarkar, "Priority-Based Time-Slot Allocation in Wireless Body Area Networks during Medical Emergency Situations: An Evolutionary Game-Theoretic Perspective," *IEEE Journal of Biomedical and Health Informatics*, vol. 19, no. 2, pp. 541–548, 2015.
[7] S. Sarkar and S. Misra, "From Micro to Nano: The Evolution of Wireless Sensor-Based Health Care," *IEEE Pulse*, vol. 7, no. 1, pp. 21–25, 2016.
[8] S. Sarkar, S. Misra, B. Bandyopadhyay, C. Chakraborty, and M. S. Obaidat, "Performance Analysis of IEEE 802.15.6 MAC Protocol under Non-Ideal Channel Conditions and Saturated Traffic Regime," *IEEE Transactions on Computers*, vol. 64, no. 10, pp. 2912–2925, 2015.
[9] S. Sarkar, S. Misra, C. Chakraborty, and M. S. Obaidat, "Analysis of reliability and throughput under saturation condition of IEEE 802.15. 6 CSMA/CA for wireless body area networks," in *Proceedings of IEEE Global Communications Conference (GLOBECOM)*, Texas, USA, 2014, pp. 2405–2410.
[10] S. Sarkar, S. Misra, and M. S. Obaidat, "Resource Allocation for Wireless Body Area Networks in Presence of Selfish Agents," in *Proceedings of IEEE Global Communications Conference (GLOBECOM)*, Washington, DC, USA, 2016, pp. 1–6.
[11] S. Sarkar, S. Chatterjee, S. Misra, and R. Kudupudi, "Privacy-Aware Blind Cloud Framework for Advanced Healthcare," *IEEE Communications Letters*, vol. 21, no. 11, pp. 2492–2495, 2017.
[12] A. Sixsmith and N. Johnson, "A Smart Sensor to Detect the Falls of the Elderly," *IEEE Pervasive Computing*, vol. 3, no. 2, pp. 42–47, 2004.
[13] L. Lu, R. Ji, and M. Liu, "Design of Real-Time Body Weight Monitor Systems Based on Smart Phones," in *Proceedings of International Conference on Mechatronics and Control*, Jinzhou, China, 2014, pp. 1392–1396.
[14] D. D. Rowlands, L. Laakso, T. McNab, and D. A. James, "Cloud Based Activity Monitoring System for Health and Sport," in *Proceedings of International Joint Conference on Neural Networks*, Brisbane, Australia, 2012, pp. 1–5.
[15] X. Masip-Bruin, E. Marin-Tordera, A. Alonso, and J. Garcia, "Fog-To-Cloud Computing (F2C): The Key Technology Enabler for Dependable E-Health Services Deployment," in *Proceedings of Mediterranean Ad Hoc Networking Workshop*, Spain, 2016, pp. 1–5.

[16] L. Gu, D. Zeng, S. Guo, A. Barnawi, and Y. Xiang, "Cost Efficient Resource Management in Fog Computing Supported Medical Cyber-Physical System," *IEEE Transactions on Emerging Topics in Computing*, vol. 5, no. 1, pp. 108–119, 2017.

[17] D. Borthakur, H. Dubey, N. Constant, L. Mahler, and K. Mankodiya, "Smart Fog: Fog Computing Framework for Unsupervised Clustering Analytics in Wearable Internet of Things," in *Proceedings of IEEE Global Conference on Signal and Information Processing*, Montreal, Canada, 2017, pp. 472–476.

[18] R. Craciunescu, A. Mihovska, M. Mihaylov, S. Kyriazakos, R. Prasad, and S. Halunga, "Implementation of Fog Computing for Reliable E-Health Applications", in *Proceedings of Asilomar Conference on Signals, Systems and Computers*, Pacific Grove, 2015, pp. 459–463.

[19] A. Pantelopoulos and N. G. Bourbakis, "A Survey on Wearable Sensor-Based Systems for Health Monitoring and Prognosis," *IEEE Transactions on Systems, Man, and Cybernetics, Part C*, vol. 40, no. 1, pp. 1–12, 2010.

[20] V. Shnayder, B. Chen, K. Lorincz, T. R. F. Fulford Jones, and M. Welsh, "Sensor Networks for Medical Care," in *Proceedings of the 3rd ACM International Conference on Embedded Networked Sensor Systems (SenSys)*, New York, 2005, pp. 314–314.

[21] G. Wu, S. Talwar, K. Johnsson, N. Himayat, and K. D. Johnson, "M2M: From Mobile to Embedded Internet," *IEEE Communications Magazine*, vol. 49, no. 4, pp. 36–43, 2011.

[22] A. Asadi, Q. Wang, and V. Mancuso, "A Survey on Device-to-Device Communication in Cellular Networks," *IEEE Communications Surveys & Tutorials*, vol. 16, no. 4, pp. 1801–1819, 2014.

[23] D. Kreutz, F. M. V. Ramos, P. E. Verissimo, C. E. Rothenberg, S. Azodolmolky, and S. Uhlig, "Software-Defined Networking: A Comprehensive Survey," in *Proceedings of the IEEE*, vol. 103, no. 1, pp. 14–76, 2015.

[24] Y. Toor, P. Muhlethaler, A. Laouiti, and A. D. La Fortelle, "Vehicle Ad Hoc Networks: Applications and Related Technical Issues," *IEEE Communications Surveys & Tutorials*, vol. 10, no. 3, pp. 74–88, 2008.

[25] M. Dikmen and C. M. Burns, "Autonomous Driving in the Real World: Experiences with Tesla Autopilot and Summon," in *Proceedings of the 8th ACM International Conference on Automotive User Interfaces and Interactive Vehicular Applications*, New York, 2016, pp. 225–228.

[26] J. Liu, J. Li, L. Zhang, F. Dai, Y. Zhang, X. Meng, and J. Shen, "Secure Intelligent Traffic Light Control Using Fog Computing," *Future Generation Computer Systems* vol. 78, no. 2, pp. 817–824, 2018.

[27] X. Hou, Y. Li, M. Chen, D. Wu, D. Jin, and S. Chen, "Vehicular Fog Computing: A Viewpoint of Vehicles as the Infrastructures," *IEEE Transactions on Vehicular Technology*, vol. 65, no. 6, pp. 3860–3873, 2016.

[28] N. B. Truong, G. M. Lee, and Y. Ghamri-Doudane, "Software Defined Networking-Based Vehicular Adhoc Network with Fog Computing," in *Proceedings of IFIP/IEEE International Symposium on Integrated Network Management*, Ottawa, Canada, 2015, pp. 1202–1207.

[29] J. Dutta and S. Roy, "IoT-fog-cloud Based Architecture for Smart City: Prototype of a Smart Building," in *Proceedings of the 7th International Conference on Cloud*

Computing, Data Science & Engineering – Confluence, Noida, India, 2017, pp. 237–242.

[30] M. Amadeo, A. Molinaro, S. Y. Paratore, A. Altomare, A. Giordano, and C. Mastroianni, "A Cloud of Things Framework for Smart Home Services Based on Information Centric Networking," in *Proceedings of the IEEE 14th International Conference on Networking, Sensing and Control*, Calabria, Italy, 2017, pp. 245–250.

[31] Y. Nikoloudakis, S. Panagiotakis, E. Markakis, E. Pallis, G. Mastorakis, C. X. Mavromoustakis, and C. Dobre, "A Fog-Based Emergency System for Smart Enhanced Living Environments," *IEEE Cloud Computing*, vol. 3, no. 6, pp. 54–62, 2016.

[32] B. R. Stojkoska and K. Trivodaliev, "Enabling Internet of Things for Smart Homes through Fog Computing," in *Proceedings of the 25th Telecommunication Forum*, Belgrade, Serbia, 2017, pp. 1–4.

[33] F. Y. Okay and S. Ozdemir, "A Secure Data Aggregation Protocol for Fog Computing Based Smart Grids," in *Proceedings of the 12th IEEE International Conference on Compatibility, Power Electronics and Power Engineering*, Doha, Qatar, 2018, pp. 1–6.

[34] F. Y. Okay and S. Ozdemir, "A Fog Computing Based Smart Grid Model," in *Proceedings of International Symposium on Networks, Computers and Communications*, Yasmine Hammamet, Tunisia, 2016, pp. 1–6.

[35] L. Lyu, K. Nandakumar, B. Rubinstein, J. Jin, J. Bedo, and M. Palaniswami, "PPFA: Privacy Preserving Fog-Enabled Aggregation in Smart Grid," *IEEE Transactions on Industrial Informatics*, vol. 14, no. 8, pp. 3733–3744, 2018.

[36] F. J. Ferrandez-Pastor, J. M. Garcia-Chamizo, M. Nieto-Hidalgo, and J. Mora-Martnez, "Precision Agriculture Design Method Using a Distributed Computing Architecture on Internet of Things Context," *Sensors*, vol. 18, no. 1731, pp. 1–21, 2018.

[37] A. Munir, P. Kansakar, and S. U. Khan, "IFCIoT: Integrated Fog Cloud IoT: A Novel Architectural Paradigm for the Future Internet of Things," *IEEE Consumer Electronics Magazine*, vol. 6, no. 3, pp. 74–82, 2017.

[38] C. Lai, D. Song, R. Hwang, and Y. Lai, "A QoS-aware Streaming Service over Fog Computing Infrastructures," in *Proceedings of Digital Media Industry & Academic Forum*, Santorini, Greece, 2016, pp. 94–98.

[39] Z. Su, Q. Xu, J. Luo, H. Pu, Y. Peng, and R. Lu, "A Secure Content Caching Scheme for Disaster Backup in Fog Computing Enabled Mobile Social Networks," *IEEE Transactions on Industrial Informatics*, vol. 14, no. 10, pp. 4579–4589, 2018.

[40] S. Park, O. Simeone, and S. S. Shitz, "Joint Optimization of Cloud and Edge Processing for Fog Radio Access Networks," *IEEE Transactions on Wireless Communications*, vol. 15, no. 11, pp. 7621–7632, 2016.

[41] S. M. Azimi, O. Simeone, A. Sengupta, and R. Tandon, "Online Edge Caching and Wireless Delivery in Fog-Aided Networks with Dynamic Content Popularity," *IEEE Journal on Selected Areas in Communications*, doi: 10.1109/JSAC.2018.2844961, 2017.

[42] B. Tang, Z. Chen, G. Hefferman, S. Pei, T. Wei, H. He, and Q. Ynag, "Incorporating Intelligence in Fog Computing for Big Data Analysis in Smart Cities," *IEEE Transactions on Industrial Informatics*, vol. 13, no. 5, pp. 2140–2150, 2017.

[43] Y. Lin and H. Shen, "Leveraging Fog to Extend Cloud Gaming for Thin-Client MMOG with High Quality of Experience," in *Proceedings of the 35th IEEE International Conference on Distributed Computing Systems*, Columbus, OH, 2015, pp. 734–735.

[44] A. Seitz, D. Henze, J. Nickles, M. Sauer, and B. Bruegge, "Augmenting the Industrial Internet of Things with Emojis," in *Proceedings of the 3rd International Conference on Fog and Mobile Edge Computing*, Barcelona, Spain, 2018, pp. 240–245.

[45] N. Chen, Y. Chen, Y. You, H. Ling, P. Liang, and R. Zimmermann, "Dynamic Urban Surveillance Video Stream Processing Using Fog Computing," in *Proceedings of the 2nd IEEE International Conference on Multimedia Big Data*, Taipei, 2016, pp. 105–112.

Chapter 10

Fog Architecture

This chapter presents the architectural details of the fog computing paradigm and elaborates its intricacies. Like most distributed computing architectures, the different architectures proposed for fog computing can be broadly classified into two categories: application agnostic and application specific. We discuss both the categories, beginning with a simplified and generic fog computing architecture and illustrating the specification and purpose of each component to give readers a coherent understanding. Based on this service architecture, we present a theoretical model to mathematically represent the different components of fog computing architecture and the inter-relation among them. In the latter part of the chapter, we discuss different advanced and service-specific architectures of fog computing. A major part of this discussion focuses on the communication architecture, resource management, and the fog-cloud interplay in these architectures.

10.1 The Comprehensive Framework

Among the pioneering works on fog computing architecture, Bonomi *et al.* [1] was one of the first to characterize the fog computing framework and justify its suitability for the services and applications of the IoT. Since then, several architectures – both application agnostic [2–9] and application specific [10–16] – have been proposed by researchers. However, before we excavate the different service architectures for fog computing, in this section, we present a simplified yet comprehensive fog computing architecture.

Fog computing is a distributed computing paradigm, and the computational nodes are envisioned to be positioned in different tiers or strata of the service architecture. Fog architecture can be envisioned as three-tiered service architecture, as shown in Figure 10.1. The composition and role of the three different tiers are discussed below.

(a) *Tier 1*: This is the bottom-most tier of the service architecture, and it comprises billions of edge devices that include personal devices owned by end-users or the different IoT devices at the network's edge. An edge device can be either mobile (a smartphones, a sensor on a car, a smartwatch, or a laptop) or static (a sensor in a parking lot or a building). They are often equipped with wireless communication modules, such as Bluetooth, ZigBee, or WiFi, in order to facilitate periodic data transmission. These devices are usually owned by the end-users; hence, they are the primary stakeholders of this tier.

(b) *Tier 2*: The middle tier, also known as the fog computing tier, can be imagined to be formed by multiple geo-spatially distributed fog instances. Each of these fog instances comprises a set of fog nodes, which can be routers, gateways, switches, access points, set-top boxes, personal computers, or even small servers. Note that, apart from their regular tasks, such as packet routing, packet forwarding, or address resolution, these devices are empowered with additional computing and storage hardware, which allow them to perform some amount of computational and decision-making tasks within this tier. The infrastructure at this tier may be owned and

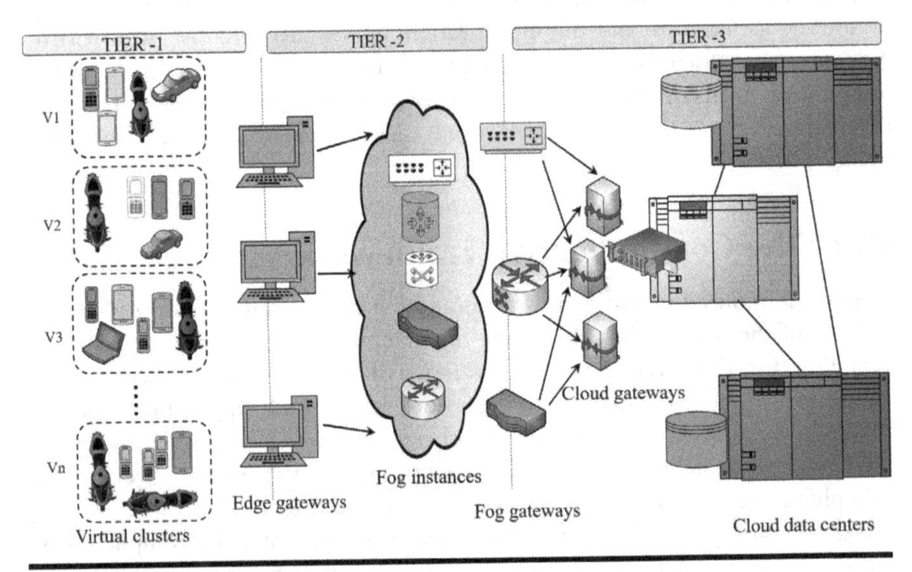

Figure 10.1 Fog computing architecture

managed by one or more Internet service providers (ISPs) or in some cases by dedicated service providers who are the main stakeholders.

(c) *Tier 3*: The uppermost tier is commonly known as the cloud-computing tier and encompasses the different geo-distributed cloud DCs. Like the classical cloud-computing framework, the DCs are powered by high-end servers and storage units. The CSPs are the primary stakeholders of the layer.

10.1.1 Communication and Network Model

As already discussed, the bottom-most tier of the service architecture comprises billions of smart, miniaturized, and Internet-connected edge devices. It is assumed that these edge devices are aware of and are able to transmit their geo-spatial locations through technologies such as the global positioning system, geographic information system (GIS), or global navigation satellite system. The edge devices, which are co-located around a geographic location, are assumed to form a location-based logical cluster, known as the virtual cluster (VC). While an edge device can be a part of only one VC at any given time, the mobility of the device allows it to switch between VCs based on its location. In other words, as VCs are location-specific logical groups of edge devices, if a mobile edge device moves from one location to another, it may have switched between VCs. This event of switching between VCs is analogous to cellular handoff for mobile phones.

However, while the data are transmitted upwards (towards the fog tier) they are processed within the intermediate fog devices. The fog devices can stretch from different networking components such as routers, switches, gateways, and access points to high-end proxy servers and computing machines. Bonomi *et al.* [3] proposed that the fog computing architecture can be subdivided into two layers: (a) the fog abstraction layer, responsible for managing the fog resources, facilitating virtualization, and maintaining privacy of the tenants, and (b) the fog orchestration layer, comprised of a small software agent named foglet, a distributed database and a service orchestration module. The foglet monitors the state of the devices; the database is used for ensuring scalability and fault-tolerance, while the service orchestration module is responsible for the policy-based routing of application requests. Data processing and analysis is carried out in the fog instances (FIs) [17,18]. Thereafter, applications that require data storage or historical data-based analytics are transmitted to the cloud DCs, and those that do not require this are processed in the fog units. The fog devices are equipped with a limited amount of semi-permanent storage capacity, which allows temporary storage of application data. Thus, they are capable of serving the delay-sensitive applications in real time.

The cloud-computing tier comprises the highly powerful cloud DCs having massive computational and storage capabilities; it is mainly responsible for the permanent storage and processing of voluminous chunks of data. However, unlike conventional cloud architecture, fog computing prevents the cloud DCs from being bombarded with every query, thereby ensuring efficient and controlled utilization of the cloud tier [19,20].

10.2 Mathematical Model of the System

Before we go into the details of the application-agnostic and application-specific fog computing architectures, in this section, we discuss the mathematical model of the generic fog computing architecture (refer to Figure 10.1) [18]. This system model aims at defining the different composite fog computing entities and the operational functions involved. In the formulation of the mathematical model, it is assumed that the total number of edge devices in the system does not change over time. However, an edge device may move across regions, which means that its belongingness to a VC may change over time, based on the device mobility. We also assume that the VCs provide complete coverage for all edge devices present in the lowest tier. In other words, all edge devices are assumed to be part of a VC all the time. We illustrate the physical significance of the assumptions later in this section. Having laid down the assumptions relating to the system architecture, we will define the different physical and virtual components of fog architecture [18], beginning with the devices at the lowest layer of the system architecture located at the edge of the network.

Definition 15. *(Edge device)* An edge device[1] (also often represented as terminal node, IoT device etc.), denoted by $\mathcal{T}$, is defined as a 6-tuple:

$$\mathcal{T} = <T_{id}, T_{st}, \tau_i, \mathcal{L}, \mathcal{H}, \mathcal{I}[q]>,$$

where, T_{id} is an alpha-numeric string representing the unique ID of the edge device.

The rest of the tuples (i.e.,T_{st}, τ_i, $\mathcal{L}$, $\mathcal{H}$, $\mathcal{I}[q]$) are used to represent the characterization of the edge device or for specification purposes and are defined as follows.

Definition 16. *(Status of an edge device)* A status of an edge device, T_{st}, defines whether the edge device is in active state and is represented as a boolean, $T_{st} = \{0, 1\}$, where the values 0 and 1 symbolize the *inactive* and *active* states, respectively.

An active device is expected to operate upholding its operational principles such as sensing of event(s), transmission of sensed information, and performing some preprocessing on the sensed data in some cases. An active device, which does not operate based on its design principles, is considered to be malfunctioning, misbehaving, or completely nonfunctioning; discussions on these types of devices are beyond the scope of this book.

Definition 17. *(Type of edge device)* The type of an edge device (τ_{i}) indicates the type of event a node senses. Mathematically, it is presented as an element of the set $\tau = \{\tau_1,\tau_2,\ldots,\tau_{\mathrm{p}}\}$, where τ denotes the set of all events being monitored by the edge devices, and p is the total number of distinct pre-listed types of event. τ_i is represented as an alpha-numeric string.

Definition 18. *(Location of an edge device)* The geo-spatial location of an edge device is defined as a 4-tuple:

$$\mathcal{L} = <l_x, l_y, l_z, t_s>,$$

where l_x, l_y, and l_z are of float data type and represent the longitude, latitude, and altitude of an edge device, respectively. t_s is of the type date and denotes the time-stamp at which the node transmits its location.

Property 1. *The belonging of an edge device in a VC at time t, is independent of its belonging at time instant* $t-1$.

The tuple $\mathcal{H}$ dictates the specifications of an edge device, which includes its hardware details, along with its mode of operation, operational frequency, and sampling rate.

Definition 19. *(Specifications of an edge device)* The specifications of an edge device ($\mathcal{H}$) is represented as a 6-tuple:

$$\mathcal{H} = <\mathcal{P}, \mathcal{M}, \mathcal{B}, \mathcal{S}, c, f>,$$

where $\mathcal{P}$ denotes the processor specifications of the node, which includes the details such as processing core speed, bus specifications, and internal register (cache memory) size as stated in [18]. The primary memory (RAM) specifications, such as memory size, memory clock, and data rate are stored in $\mathcal{M}$. The tuple $\mathcal{B}$ describes the battery details (viz., voltage, size (AA or AAA), type (Ni or C electrodes), and the number of battery cells required). $\mathcal{S}$ is the symbolic representation of the different types of sensors that are used as sub-modules of the node. The tuple c denotes the hardware used for wireless communication for the node, such as Bluetooth and ZigBee. The frequency range in which the edge device operates is denoted by f.

The last tuple in the definition of an edge device, $\mathcal{I}[q]$, is a linear data structure, such as a 1-D array (with q elements) that stores the instance IDs of the application instances running on the device. Although applications have independent existences, it is impossible for an application instance to exist without the presence of a parent device.

Having defined the tuples associated with an edge device, we now define the terms *application* and *application instance* accordingly [18]. The need for separating an application from an application instance is triggered by the fact that there can be multiple instances of the same application running within a single edge device concurrently. This may be analogously compared with the notion of *class* and *object* in an object-oriented programming language. We know that an object is simply an instance of a class and there can be multiple objects of the same class interacting among themselves in a program. Similarly, there can several copies of the same application running with a single edge device. Mathematically, we define these two terms as follows:

Definition 20. *(Application)* An application $\mathcal{A}$ is defined as a 4-tuple:

$$\mathcal{A} = <A_{id}, A_{type}, A_{sp}>,$$

where A_{id} is the application ID and A_{type} denotes the purpose for which the application is used (medical, education, finance, entertainment, utility, and gaming, etc.). A_{sp} dictates the minimum system specifications required to run the application including the processor, primary memory, secondary storage details, and the operating system version.

Definition 21. *(Application instance)* The instance of an application, $\mathcal{I}$, is defined as a 5-tuple:

$$\mathcal{I} = <I_{id}, A_{id}, T_{id}, I_{st}, I_{req}>,$$

where I_{id} is the application instance ID that can be thought of as the process ID generated by the system. A_{id} and T_{id} bear the same meanings as mentioned earlier. I_{st} is a boolean variable, such that $I_{st} = \{0, 1\}$, where the values 0 and 1 are indicative of the application instance being idle (not inactive) and active, respectively. The final tuple I_{req} is the resource requirement of the application instance. This resource requirement may be in terms of network bandwidth (for streaming applications), computation and analysis ability (for medical applications), or storage and processing power (for gaming applications). Multiple instances of an application may concurrently run on an edge device; they are distinguished by their unique instance IDs.

Next, we define the virtual or logical component of this layer, which is the VC [2]. As mentioned before, a VC is simply a logical boundary that represents all edge devices located within the physical region corresponding to that boundary.

Definition 22. *(Virtual cluster or VC)* A VC, denoted by $\mathcal{V}$, corresponds to a logical boundary comprising several localized edge devices and is mathematically defined as a 4-tuple:

$$\mathcal{V} = <V_{id}, \mathcal{T}^t[u], R, F_{id}>,$$

where V_{id} is the ID of the VC, and $\mathcal{T}^t[u]$is a non-empty 1-D array of size u that stores the IDs of all the constituent edge devices at time instant, t. As the edge devices may be mobile and move from one physical region to another, the set of such devices within a VC must strictly be represented as a function of time. Thus, it is evident that the array should have a dynamic length, where the array length, at any given time instant, is indicative of the number of edge device that are present in the VC, and the length changes as some mobile edge devices leave or join the cluster. The physical region monitored by a VC, which encloses all the intermediate edge devices, is denoted by R. R represents the geo-spatial location of a set of points, which constitutes the perimeter of the region. Mathematically, R is represented as $R = r_1, r_2, \ldots, r_u$, where each r_i corresponds to a single point on the boundary of the region. Once again, much like $\mathcal{L}$, every r_i is represented as a three-tuple: $r_i = <r_{ix}, r_{iy}, r_{iz}>$, which essentially dictates the longitude, latitude, and altitude of the physical point. For efficient monitoring of all the edge devices, we have divided the geographic terrain into multiple non-overlapping regions – each region mapped to a VC. The physical FI to which a VC is mapped is referred to by the fog instance ID, F_{id}.

Having looked at all the essential components of tier 1, we move up to the middle tier, the fog computing tier. We now define the term fog instance and the composite fog computing modules, which together constitute the middle tier [18].

Definition 23. *(Fog instance)* A fog computing instance, or simply an FI, denoted by ($\mathcal{F}$), is mathematically defined as a 3-tuple:

$$\mathcal{F} = <F_{id}, C_{AP}, \mathcal{D}[v]>,$$

where F_{id} is the unique ID of the fog computing instance, and C_{AP} is the access point through which the FI is connected to the core cloud computing framework. The third tuple, $\mathcal{D}[v]$, is a non-empty 1-D array of size v, which stores the device IDs of all the constituent fog computing devices of a FI.

Property 2. *The mapping from the set of VCs to the set of FIs, represented as* $f(\cdot) : \tilde{\mathcal{V}} \rightarrow \tilde{\mathcal{F}}$, *is injective.*

Property 3. *The mapping from the set of edge devices to the set of FIs, denoted as* $f'(\cdot) : \tilde{\mathcal{T}} \rightarrow \tilde{\mathcal{F}}$, *is many-to-one.*

Definition 24. *(Fog computing device)* We define a fog computing device, $\mathcal{D}$, in terms of its type and characteristic features in the form of a 3-tuple:

$$\mathcal{D} = < D_{id}, D_{type}, D_{sp} >$$

where, D_{id} and D_{type} are the ID and the type (such as gateway, router, processing unit, or storage) of the fog computing device. The hardware and related specifications of the device are stored in the D_{sp} tuple.

Now that we have looked at the definitions of all components of fog computing architecture, we will establish the relationships among the different entities.

Proposition 10.1. *The pairwise intersection of the VCs is null.*

Proof. We prove this by the method of contradiction. We assume, $\exists \mathcal{V}_i, \mathcal{V}_j$, such that $\mathcal{V}_i \bigcap \mathcal{V}_j \neq \Phi$. Thus,

$\exists \mathcal{T}_k$ such that,

$$\mathcal{T}_k \in \mathcal{V}_i, \mathcal{V}_j \Rightarrow \mathcal{T}_k \in \mathcal{V}_i \bigcap \mathcal{V}_j \tag{10.1}$$

As per Property 2,

$$f'(\mathcal{T}_k) = \mathcal{F}_p, \mathcal{F}_p \in \tilde{\mathcal{F}} \Rightarrow f^{-1}(\mathcal{F}_p) = \{\mathcal{V}_i, \mathcal{V}_j\} \tag{10.2}$$

which is absurd, as per injectivity of Property 1. Thus, our assumption is invalid. This concludes the proof.

Proposition 10.2. *At any given time instant,* $|\tilde{\mathcal{T}}| = |\mathcal{V}_1| + |\mathcal{V}_2| + \ldots + |\mathcal{V}_m|$, *where* $|\tilde{\mathcal{T}}|$*denotes the total number of edge devices present at tier 1.* $|\mathcal{V}_i|$ *denotes the number of edge devices mapped to the* i^{th} *VC, and m is the total number of VCs present in the system.*

Proof. Proof is done using the method of contradiction. We assume,

$$|\tilde{\mathcal{T}}| \neq |\mathcal{V}_1| + |\mathcal{V}_2| + \ldots + |\mathcal{V}_m| \tag{10.3}$$

This implies that there exists at least a pair,$(\mathcal{V}_i, \mathcal{V}_j)$, such that

$$\mathcal{V}_i, \mathcal{V}_j \in \tilde{\mathcal{V}}, \mathcal{V}_i \bigcap \mathcal{V}_j \neq \Phi, \tag{10.4}$$

which contradicts with Proposition 10.1 and disapproves our assumption. This concludes the proof.

Proposition 10.3. *The mapping $g(\cdot)$ of the set of all edge devices to the set of VCs is surjective.*

Proof. As per Properties 1 and 2, it is intuitive that $\forall \mathcal{T}_i \in \tilde{\mathcal{T}}, \exists \mathcal{V}_j \in \tilde{\mathcal{V}}$ such that $g(\mathcal{T}_i) = \mathcal{V}_j$. Therefore, $\forall \mathcal{V}_j \in \tilde{\mathcal{V}}, g^{-1}(\mathcal{V}_j) = K \subseteq \tilde{\mathcal{T}}$. If $g^{-1}(\mathcal{V}_j) = \Phi$, the length of $\mathcal{V}_j \succ \mathcal{T}[u]$ is 0. However, as per Definition 22, $\mathcal{T}[u]$ is non-empty. Thus, for every $\mathcal{V}_j$ there exists at least a single pre-image $\mathcal{T}_i$. This concludes the proof of surjectivity for $g(\cdot)$.

10.3 Application Agnostic Fog Architectures

Now that we have developed an understanding of the generic architecture of the fog computing paradigm, we are well-suited to delve into further intricacies. As we have mentioned in the introduction to the chapter, fog architectures can be principally classified into two categories: (a) application-agnostic architectures and (b) application-specific architectures. We will now discuss in detail the former.

Application-agnostic fog architecture has primarily evolved as an optimized computing infrastructure based on the modularity of end-to-end services. The key idea behind this modularity is to split the application into a set of inter-related yet independent services or tasks. These tasks may be executed sequentially or in parallel. In either case, the output of the final task or set of tasks is exactly the same as that of the case when the application is executed as a whole. It is also often envisioned to be highly aligned to the concept of dividing the life cycle of the application into different phases, such as development, deployment, execution, and management. Once the different phases of the application life cycle are visited sequentially or iteratively, the expected result is obtained. This model or application execution and service provisioning is particularly relevant with the growth of distributed computing architectures, such as cloud and cluster computing. For similar reasons, it holds a high relevance in the context of the fog computing paradigm.

In fog computing, services are provisioned from both the fog layer and the cloud layer, based on the service type, requirements, and specification. For example, while most latency-sensitive applications are served from near the edge (i.e., from the fog computing layer), services that require historical data processing or global data analysis are directed to the cloud computing core. Therefore, there is need for a classifier capable of analyzing the service requirements and, based on that, determining their service location. This classifier, often referred to as the orchestrator, is responsible for determining the service location of an application and orchestrating between the fog and cloud computing layers. For modular classification of these services, in fog computing, we envisage dedicated middleware, which may or may not be connected with the orchestrator unit. Based on the works of Issarny *et al.* [2] and Teixeira *et al.* [3], we envision a fog-based service orientation performed by this middleware at this intermediate layer. This essentially breaks the service into several micro-services, whenever there is a service requested from the bottom layer. The architecture proposed comprises three primary components: service producers, service consumers, and a service registry, as shown in Figure 10.2. The service consumers are actually the edge devices requesting certain services, while the service producers correspond to the service providers at the computing units providing the services. Any new service offered by the service producers is first registered in the service registry. After an agreement is signed between the concerned service providers and the end users, all service requests generated from the corresponding service consumer units are directly fetched from the service registry. All necessary communication between the different pairs of the components takes place over the Internet.

This model represents the typical application-agnostic service architecture. In the fog computing service model, we typically have two service providers – fog service provider and cloud service provider – each providing a separate set of services. Of course, in some special cases, the fog and cloud service providers are the same entity; currently, we discuss the generic service model with different stakeholders governing the infrastructures at the different layers. We present the framework for the application-agnostic fog computing architecture, where multiple fog and cloud service providers (stakeholders) are operational at different layers of the architecture. As shown in Figure 10.3, there are two cloud service providers (X_1 and X_2) and three fog service providers (Y_1, Y_2, and Y_3). Each fog service serves a number of edge devices. Y_1 and Y_2 are both connected with X_1 and X_2. This means that the edge devices connected to either Y_1 or Y_2 are allowed to choose to obtain any service from a set of cloud services offered by X_1 and X_2. Y_3, however, has connection only with X_2; thus, the edge devices connected with Y_3 may choose their services from the restricted set of services provided by X_2. Although this is not a popular fog use case, it gives us a perspective on how a small-scale and private fog infrastructure may be managed.

10.3.1 Programming Models

In recent times, a few programming models have been proposed by the researchers focusing particularly on the IoT and supported by application-agnostic fog architectures. One of the earliest and now highly popular among them is the

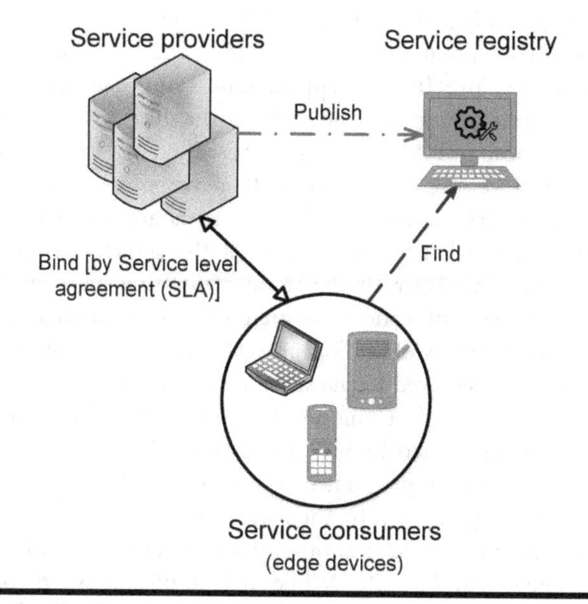

Figure 10.2 Application agnostic service architecture [18]

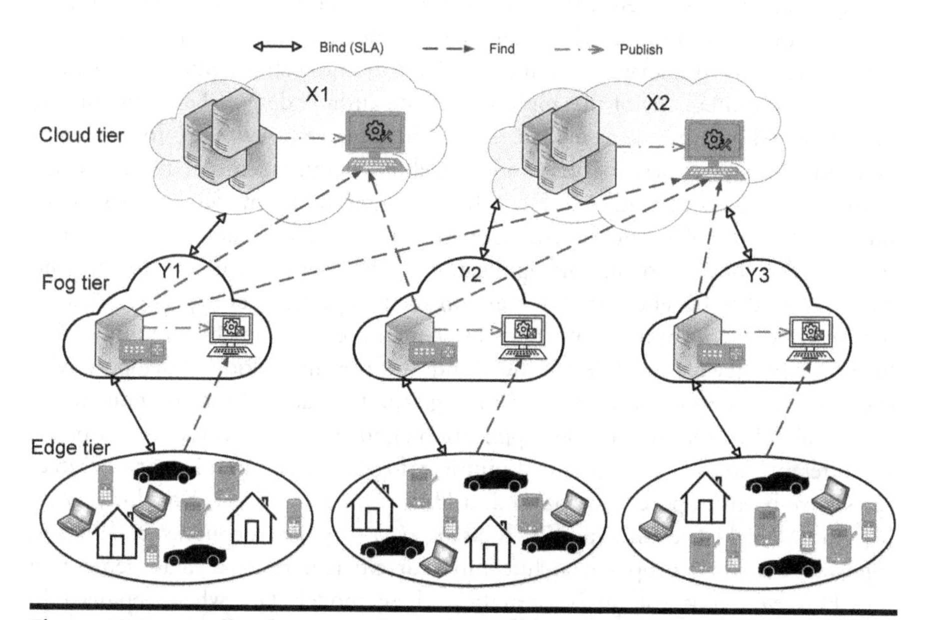

Figure 10.3 Application agnostic service architecture for fog computing [3]

Mobile Fog programming model proposed by Hong *et al.* [21]. Mobile Fog is essentially a high-level programming model for large-scale, geo-spatially distributed, and latency-sensitive Internet applications – the same class of application that remains the primary motivation behind the birth of the fog computing paradigm. The concept of Mobile Fog is based on the generic three-tier fog computing architecture (see Figure 10.1). It allows programmers to write programs for specific nodes in each tier. Programmers are allowed to program the edge/IoT devices at the bottom-most layer of the service architecture, the fog nodes at the intermediate layer, or the cloud computing nodes at the top of the hierarchy. Once a piece of code is written by a programmer, the developers compile it and generate a Mobile Fog process image against it. This image can then be uploaded and deployed to the specific node for which it is targeted. Each such image is associated with a unique identifier, and deployment of the image can be done in devices located in any of the tiers. Mobile Fog also provides the programmers with a set of programming interfaces, allowing them to control applications by referring to them by their identifier.

Another programming model was proposed by Giang *et al.*, which is based on distributed dataflow (DDF) [22]. While IoT applications are often typically characterized by the device heterogeneity and requirement for device mobility, they are also expected to support a large number of service requests generated by billions of edge devices. The most important characteristic of these applications, however, concerns the latency-sensitivity towards the service, known as the perception-action cycle. As shown in Figure 10.4, the perception-action cycle is governed by the communication among pairs of devices located at different layers. The challenge is to manage these inter-device communications efficiently to meet the service requirements of the application. The authors designed a programming model by taking into account the device heterogeneity, support for different perception-action cycles, mobility, and scalability requirements of the fog computing and IoT applications. The DDF programming model, in this light, provides an efficient means for the development of fog applications and manages resource distribution to support applications. The two types of developers who are envisioned to benefit from DDF are (a) the component developers responsible for design and development of the components to ensure timely communication between the devices at different tiers and (b) the application developers who create the scripting components and manage the flow among the components.

Several other specific-purpose application-agnostic architectures are proposed by different authors. For instance, multiple fog-based communication architectures have been proposed by Aazam and Huh [4], Shi *et al.* [5], Khrishnan *et al.* [6], and Slabicki and Grochla [7]. Kapsalis *et al.* [8], Bittencourt *et al.* [9], and Agarwal *et al.* [23] proposed architectures for efficient resource management in fog. However, it is still to be seen how these models fare when deployed in practice.

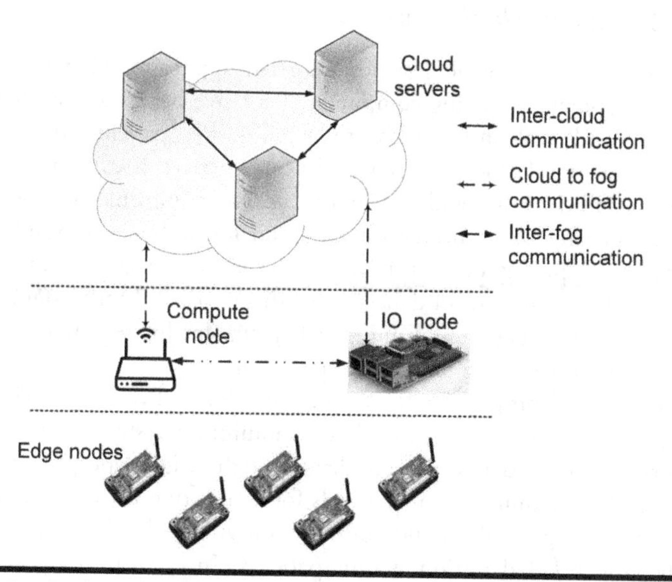

Figure 10.4 Communications in perception-action cycle [22]

10.4 Application-Specific Fog Architectures

Development of application-specific fog architecture has gained popularity recently. Unlike the application-agnostic architectures, application specific architectures are based on the application requirements, and the processing locations (i.e., at the fog nodes or some at the cloud core) of the different parts of the application are determined. While these architectures are designed to serve certain applications efficiently, due to their application-specific design and tuning, the modularity property of the application is absent. Rather, for application-specific fog applications, the orchestrator plays a critical role in determining the processing location of the different variants of the applications and managing the service requests accordingly. We present few key fog architectures designed to serve specific applications.

10.4.1 Fog Architectures for Healthcare

While the spectrum of healthcare services includes a wide range of use cases, such as pervasive and ubiquitous healthcare, activity and health monitoring, ECG and EEG feature extraction, and assisting patients with different medical conditions, most application-specific fog architectures are directed to a specific use case. We discuss some of the main fog architectures facilitating some important medical use cases.

10.4.1.1 Pervasive Healthcare

Pervasive healthcare is one of the most targeted application domains for infrastructure development using fog computing. Pervasive healthcare facilitates seamless, remote, and real-time healthcare services for patients from a global perspective. The significance of such healthcare services is extremely important and useful for fall detection and continuous health monitoring of elderly people, for patients suffering from chronic diseases, and for ambulatory healthcare [24]. The fundamental principles of this mode of healthcare are sensor-based data acquisition, real-time analysis of data, remote diagnosis, and early commencement of medical treatment. Fog computing is a key enabler for real-time data processing and service provisioning for these applications.

Cao *et al.* [10,25] proposed a fog architecture focused on the development of edge computing-based real-time healthcare monitoring system. The architecture of this work comprises data-capturing devices at the edge (laptops, smartphones, etc.), inter-device communication channels for connecting the edge devices to the back end, and storage analysis and computation at the cloud servers. This work specifically targets fall detection within patients. In the case of positive event detection, an alarm is triggered by the fog computing nodes. However, the data generated by the device is subsequently redirected to the cloud servers as well for further analysis of data for event classification and global data analysis.

Stanchev *et al.* [11] proposed fog architecture for activity and health monitoring, as shown in Figure 10.5. The authors proposed a process-oriented view of the healthcare application and modeled it using business process-modeling notation. The architecture proposed comprises three tiers, with the data generating units (i.e., physiological sensor nodes) located at the bottom-most tier. In the upper tier, the fog gateways are envisioned to act as the fog computing nodes and are responsible for short-term storage of patients' physiological data and for performing preliminary-level processing. At the top of the hierarchy is the cloud computing core, which is responsible for permanent storage and access control of the data. Although this architecture utilizes the fog stratum for short-term storage and pre-processing of data, the fact that the doctors will have to fetch data from the cloud data centers while processing may incur some undesired latency.

In some cases, researchers suggested a four-tier fog architecture to support application-specific service provisioning. Nandyala and Kim [26] proposed a four-tier fog and IoT-based real time healthcare monitoring architecture for smart home and hospital settings. As shown in Figure 10.6, at the lowest tier of the architecture are the smart edge and IoT devices, which are responsible for capturing physiological data from patients and sending them to the gateway nodes at the upper tier. This process of data acquisition relies on machine-to-machine (M2M) communication among the edge devices for the purpose of collection and pre-processing of data. The gateway nodes at the upper tier act as the fog computing nodes and are

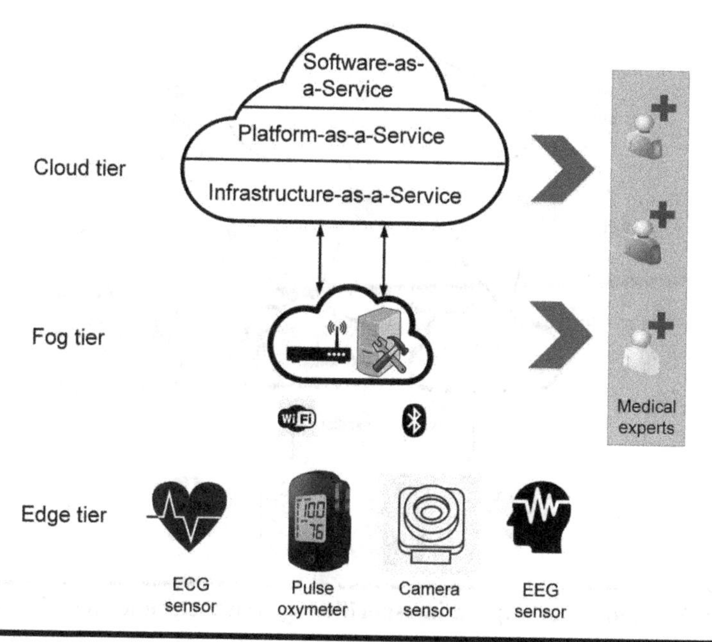

Figure 10.5 Application-specific fog architecture for activity and health monitoring [11]

responsible for serving the latency-sensitive applications locally. The tier above this is the core network tier, which provides paths for transferring data and network-related information among several sub-networks. The cloud computing tier rests at the top of the hierarchy and supports the architecture by facilitating global analytics and long-term data storage. A similar four-tier architecture for patient monitoring has been proposed by Chakraborty *et al.* [27].

10.4.1.2 ECG and EEG Feature Extraction

While the ECG or electrocardiogram is a technique for analyzing the electrical activity of the heart, EEG or electroencephalogram is a method for analyzing brain activity. Both of these techniques are widely used for health monitoring of patients, and the devices are often critical constituents of different wireless body area networks (WBANs).

A fog-based healthcare-monitoring framework, focusing specifically on ECG feature extraction was proposed by Gia *et al.* [12]. The solution captures ECG data streams generated by the ECG sensors and processes the data in real time within smart gateway nodes. The gateway nodes are equipped with the additional ability to process, mine, analyze, and store the data. Clearly, these nodes act as the fog computing nodes in this

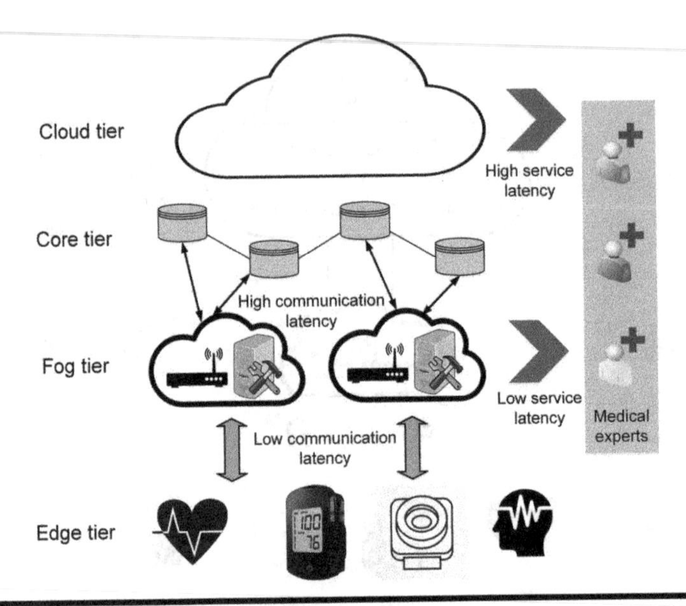

Figure 10.6 A four-tier application specific fog architecture for healthcare [26]

service architecture. This fog computing service layer contains components for supporting heterogeneity of the edge devices, inter-operability within these devices and the fog servers, and location awareness. This facilitates extraction of data with variable storage formats and helps in providing ubiquitous and location-aware services to the patients, particularly in medical emergencies.

For EEG feature extraction purposes, Zao *et al.* [28] proposed a fog-based brain-computer interface game called "EEG Tractor Beam". The game runs within the smartphones of the end-users, and every end-user is required to wear a EEG-based headset. The raw EEG data streams captured by these headsets are transmitted via smartphone to the respective local fog nodes. The fog nodes run within the application-specific components such as brain state classification and model fitting. This example demonstrates the applicability of fog computing in an area like gaming, where real-time, latency-sensitive services play a highly critical role.

10.4.1.3 Assisting Patients with Medical Conditions

Application-specific fog architectures play an important role in providing assistance to patients suffering from different medical conditions such as Parkinson's disease, dementia, and speech disorders. Several such architectures have been proposed by different research groups recently, among which we discuss a few use cases.

A fog-based interface, called FIT (short for Fog-driven IoT) was proposed by Monteiro *et al.* [29] to assist patients suffering from Parkinson's disease and speech disorders. For this, they proposed a three-tier fog architecture. At the tier closest to the network's edge are the patients. An Android-based smartwatch is used to collect the clinical speech data from the patients. These data are then transmitted by the watch to the fog tier by means of wireless communication. At the fog tier, specific applications for analyzing and interpreting the speech are deployed and real-time output based on the analysis is provided. The fog devices they used are the Intel Edison-powered computing nodes. The features extracted at the fog tier are then sent to the cloud tier on a periodic basis to facilitate long-term analysis and storage. A similar Intel Edison-based architecture for assisting patients with speech disorders was proposed by Dubey *et al.* [30].

For assisting patients suffering from mild dementia, Fratu *et al.* [13] proposed a fog computing based system. At the edge tier, several sensor nodes containing infrared-based movement sensors and temperature sensors are deployed within the patient's residence, along with the physiological sensors, which the patient is made to wear. The data collected from these sensors are then processed in real time at the fog tier with the help of task-specific application components. This system can also help patients suffering from chronic obstructive pulmonary disease (COPD).

10.4.2 Fog Architectures for Smart City Environments

Having discussed the main application specific architectures for fog computing in the domain of healthcare, we now highlight its importance with respect to smart city environments.

10.4.2.1 Smart Vehicular System

One of most extensive uses of application-specific fog architectures has been made in the domains of smart traffic management, transportation system, and vehicular technologies. Datta *et al.* [31] proposed a three-tier architecture that provides three consumer-centric services (M2M data analytics with semantics, discovery, and management of the connected vehicles). Each vehicle is equipped with multiple sensor modules. Together these sensor-equipped vehicles comprise the set of edge devices. The data sensed by these sensors in the vehicles are sent to the roadside units (RSUs) and M2M gateways, which act as the fog computing nodes. These RSUs and gateways are empowered with computing and short-term storage facilities, which allow them to analyze the data in real time and provide the three aforementioned services based on the analysis. Real-time service provisioning is a key requirement in vehicular systems. In the systems, all real-time and latency-sensitive services are provided by the nodes within the fog tier. At the cloud tier, further analysis and processing of these data are performed for services that are more interested in the global view of the system and do not come with a strict latency requirement.

In this context, the concept of vehicular fog computing has recently gained a lot of popularity. The main challenges that vehicular fog computing face are due to (i) highly dynamic edge devices moving at a high velocity, (ii) the requirement of a high degree of accuracy and precision in decision making, and (iii) quick or instantaneous recovery from failures. Bitam *et al.* [14] proposed a three-tier infrastructure, known as VANET-cloud, to improve traffic safety and real-time computational services to the mobile vehicles. The edge layer is referred to as the client layer, where the vehicular sensors and other sensory devices reside. Data collected by these edge nodes are sent over to the communication devices and RSUs at the communication layer. While all intermediate and latency-sensitive computations are performed at this tier, the cloud is used for long-term storage and global data processing. Another VANET-based traffic management architecture, as shown in Figure 10.7, was proposed by Truong *et al.* [32]. It leverages the fog and software defined networking (SDN) to achieve its objectives. Many such fog-based application-specific architectures [15,33–35] were proposed in recent times to facilitate smart vehicle and traffic management, which mostly follow a similar architectural hierarchy.

10.4.2.2 Smart Living

Smart city applications often require services at a much larger scale, given the huge number of devices at the edge of the network. Tang *et al.* [36] proposed a hierarchically distributed fog architecture that spans across four service tiers. As usual, the edge and IoT devices are located at the bottom-most tier and the cloud computing core rests at the top-most tier. Between these two, the authors proposed two more tiers where the fog computing nodes are spread. Just above the edge tier is the fog tier with many low-power and high-performance nodes. These nodes are responsible for quick processing of the collected data and timely service provisioning. On top of this tier is another group of fog computing nodes, designated to handle emergency and hazardous situations. The purpose of having the fog computing layer split into two is to prevent the network and service model from breaking down in cases of sudden spikes in service requests.

Another smart environment architecture is proposed by Li *et al.* [37]. They split the fog computing nodes into two categories: (a) fog edge nodes and (b) fog server nodes. While the fog server nodes are responsible for the management of application deployment, network configuration, and pricing based on resource utilization, the fog edge nodes are designed to provide support for computing and

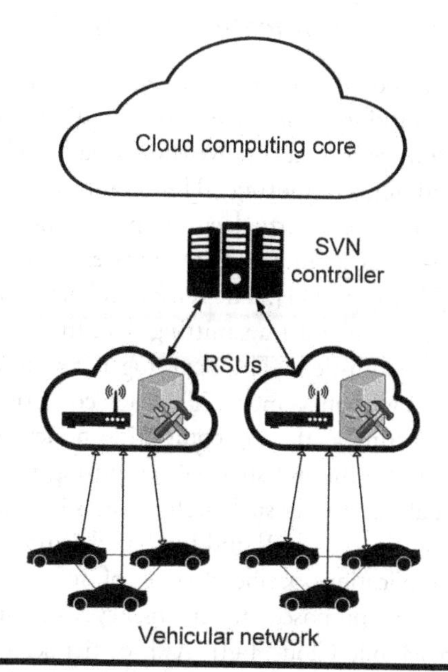

Figure 10.7 VANET-based fog architecture for smart traffic management [32]

storage services based on the server's instruction. Splitting the fog operations between these two sets of nodes, once again, helps simultaneously manage and serve a huge number of applications.

10.4.2.3 Smart Energy Management

Smart energy management and distribution are key factors in the context of smart cities. One way to facilitate these is through the deployment of a smart grid. Yan and Su [16] proposed a fog-based smart grid-driven ecosystem. Electronic devices, which consume the power distributed by the smart grids, are the edge devices. The fog nodes, in this case, are the smart meters against every consumer. A similar architecture was proposed by Okay and Ozdemir [38]; it is shown in Figure 10.8. Another fog-assisted architecture for home energy management was studied by Vatanparvar and Al Faruque [39], who demonstrated support for easy deployment, scalability, and interoperability.

10.4.3 Other Application-Specific Fog Architectures

There are many other use cases that facilitate application-specific fog computing architectures. One such use case is fog-based communication and content dissemination. We present this use case and some others.

10.4.3.1 Advanced Communication Technologies

Abdullahi *et al.* [40] proposed a different type of fog architecture; it presents information-centric networking (ICN) as API to ubiquitous computing. As shown in Figure 10.9, the authors proposed to solve the off-path caching by exploiting the advantages of ICN and fog computing. The cloud computing tier is used to optimize the discovery of services, virtualization, processing, and storage.

Another type of fog-based architecture for wireless sensor and actuation networks was proposed by Lee *et al.* [41]. The sensor and actuator nodes comprise the edge layer in this case, sensing and transmitting data to the fog tier. The fog tier, however, is split into two sub-tiers. The slave fog tier is responsible for resource management and data-flow management. This tier comprises the gateway nodes and micro-server nodes acting as the fog computing nodes. The master fog tier is empowered with more powerful and smarter computing nodes and is responsible for the control functionalities. Other such architectures include the SDN-based fog architecture proposed by Xu *et al.* [42] and fog-based radio access networks [43].

Among the other application-specific designs of fog computing-based service architectures, Aazam *et al.* proposed an emergency assistance system called the Emergency Help Alert Mobile Cloud [44]. A three-tier security fog-based security framework is suggested in [45]. Also, a drone-assisted smart surveillance architecture is proposed by Chen *et al.* [46]. In their proposed architecture, personal devices, such as smartphones and laptops, can be used as fog computing nodes to track down a surveillance target with support from the cloud for long-term data storage.

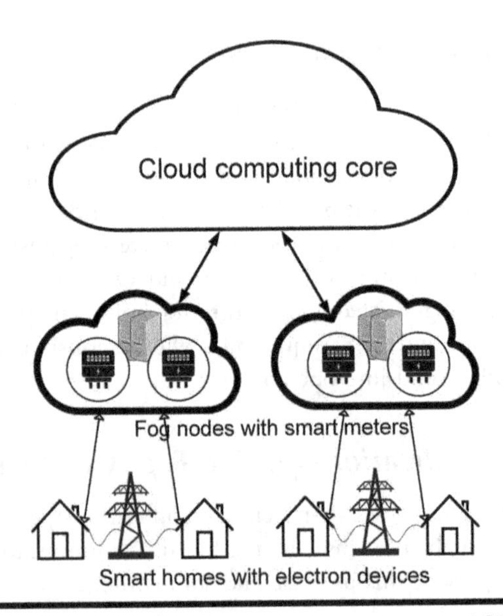

Figure 10.8 Fog architecture for smart energy management [38]

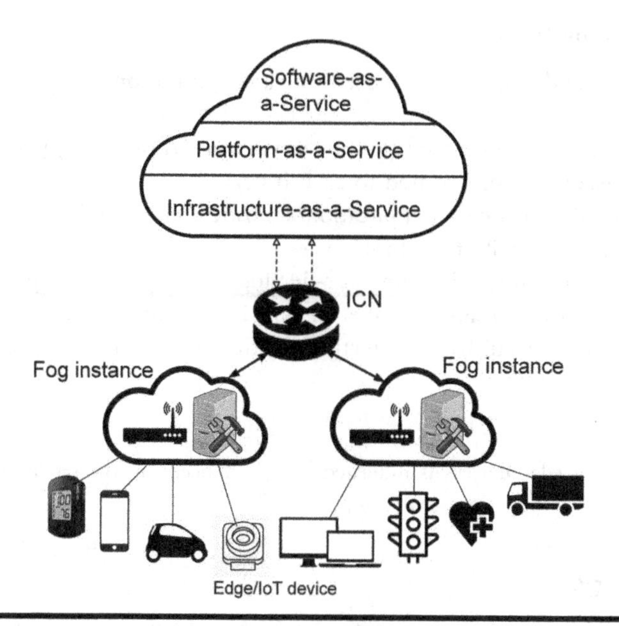

Figure 10.9 Application specific fog architecture for supporting ICN [40]

10.5 Summary

In this chapter, we started with an explanation of a comprehensive fog computing framework and presented the generic communication and network model for fog computing. Based on this model, we then discussed the mathematical characterization of the fog computing paradigm by defining its key components and their inter-relationships. Subsequently, we discussed two types of fog architectures: application-agnostic and application-specific. We also illustrated different use cases that facilitate the application-specific design of the fog architectures.

Working Exercises

Numerical Problems

1. How would you prove that the pairwise intersection of the virtual clusters is null?
2. If there are 10 billion edge devices spread homogeneously across the planet and there are 1,000 equi-sized virtual clusters, then what will be the cardinality of each virtual cluster?
3. Prove that the mapping of the set of all edge devices to the set of virtual clusters is surjective.

Conceptual Questions

1. Define and differentiate between an application and an application instance.
2. What do you understand by the terms *virtual cluster* and *fog instance*? How are these two terms related to each other?
3. Distinguish between application-agnostic and application-specific fog architectures with an example.
4. How do the three-tier and four-tier fog architectures compare? What are your views on variable number of tiers in fog architecture?
5. Describe in detail fog architecture for the smart vehicular system.

Note

1 The set of edge devices is indicated by $\tilde{T}$. Henceforth, the set of entities is thus denoted.

References

[1] F. Bonomi, R. Milito, J. Zhu, and S. Addepalli, "Fog Computing and Its Role in the Internet of Things," in *Proceedings of the 1st Edition of the MCC Workshop on Mobile Cloud Computing*, Helsinki, Finland, Aug. 2012, pp. 13–16.

[2] V. Issarny, N. Georgantas, S. Hachem, A. Zarras, P. Vassiliadist, M. Autili, M. A. Gerosa, and A. B. Hamida, "Service-Oriented Middleware for the Future Internet: State of the Art and Research Directions," *Springer Journal of Internet Services and Applications*, vol. 2, no. 1, pp. 23–45, 2011.

[3] T. Teixeira, S. Hachem, V. Issarny, and N. Georgantas, "Service Oriented Middleware for the Internet of Things: A Perspective," in *Proceedings of European Conference on a Service-Based Internet*, 2011, pp. 220–229.

[4] M. Aazam and E.-N. Huh, "Fog Computing and Smart Gateway Based Communication for Cloud of Things," in *Proceedings of International Conference on Future Internet of Things and Cloud*, 2014, pp. 464–470.

[5] H. Shi, N. Chen, and R. Deters, "Combining Mobile and Fog Computing: Using CoAP to Link Mobile Device Clouds with Fog Computing," in *Proceedings of IEEE International Conference on Data Science and Data Intensive Systems*, 2015, pp. 564–571.

[6] Y. N. Krishnan, C. N. Bhagwat, and A. P. Utpat, "Fog Computing – Network Based Cloud Computing," in *Proceedings of the 2nd International Conference on Electronics and Communication Systems*, 2015, pp. 250–251.

[7] M. Slabicki and K. Grochla, "Performance Evaluation of CoAP, SNMP and NETCONF Protocols in Fog Computing Architecture," in *Proceedings of IEEE/IFIP Network Operations and Management Symposium*, 2016, pp. 1315–1319.

[8] A. Kapsalis, P. Kasnesis, I. S. Venieris, D. I. Kaklamani, and C. Z. Patrikakis, "A Cooperative Fog Approach for Effective Workload Balancing," *IEEE Cloud Computing*, vol. 4, no. 2, pp. 36–45, 2017.

[9] L. F. Bittencourt, M. M. Lopes, I. Petri, and O. F. Rana, "Towards Virtual Machine Migration in Fog Computing," in *Proceedings of the 10th International Conference on P2P, Parallel, Grid, Cloud and Internet Computing*, 2015, pp. 1–8.

[10] Y. Cao, S. Chen, P. Hou, and D. Brown, "FAST: A Fog Computing Assisted Distributed Analytics System to Monitor Fall for Stroke Mitigation," in *Proceedings of IEEE International Conference on Networking, Architecture and Storage*, 2015, pp. 2–11.

[11] V. Stantchev, A. Barnawi, S. Ghulam, J. Schubert, and G. Tamm, "Smart Items, Fog and Cloud Computing as Enablers of Servitization in Healthcare," *Sensors & Transducers*, vol. 185, no. 2, p. 121, 2015.

[12] T. N. Gia, M. Jiang, A. M. Rahmani, T. Westerlund, P. Liljeberg, and H. Tenhunen, "Fog Computing in Healthcare Internet of Things: A Case Study on ECG Feature Extraction," in *Proceedings of IEEE International Conference on Computer and Information Technology; Ubiquitous Computing and Communications; Dependable, Autonomic and Secure Computing; Pervasive Intelligence and Computing*, 2015, pp. 356–363.

[13] O. Fratu, C. Pena, R. Craciunescu, and S. Halunga, "Fog Computing System for Monitoring Mild Dementia and COPD Patients – Romanian Case Study," in *Proceedings of the 12th International Conference on Telecommunication in Modern Satellite, Cable and Broadcasting Services*, 2015, pp. 123–128.

[14] S. Bitam, A. Mellouk, and S. Zeadally, "VANET-Cloud: A Generic Cloud Computing Model for Vehicular Ad Hoc Networks," *IEEE Wireless Communications*, vol. 22, no. 1, pp. 96–102, 2015.

[15] X. Hou, Y. Li, M. Chen, D. Wu, D. Jin, and S. Chen, "Vehicular Fog Computing: A Viewpoint of Vehicles as the Infrastructures," *IEEE Transactions on Vehicular Technology*, vol. 65, no. 6, pp. 3860–3873, 2016.

[16] Y. Yan and W. Su, "A Fog Computing Solution for Advanced Metering Infrastructure," in *Proceedings of IEEE/PES Transmission and Distribution Conference and Exposition*, 2016, pp. 1–4.

[17] S. Sarkar, S. Chatterjee, and S. Misra, "Assessment of the Suitability of Fog Computing in the Context of Internet of Things," *IEEE Transactions on Cloud Computing*, vol. 6, no. 1, pp. 46–59, 2018.

[18] S. Sarkar and S. Misra, "Theoretical Modeling of Fog Computing: A Green Computing Paradigm to Support IoT Applications," *IET Networks*, vol. 5, no. 2, pp. 23–29, 2016.

[19] T. Nishio, R. Shinkuma, T. Takahashi, and N. B. Mandayam, "Service-Oriented Heterogeneous Resource Sharing for Optimizing Service Latency in Mobile Cloud," in *Proceedings of the 1st International Workshop on Mobile Cloud Computing and Networking*, Bangalore, India, Jul. 2013, pp. 19–26.

[20] A. Seal and A. Mukherjee, "On the Emerging Coexistence of Edge, Fog and Cloud Computing Paradigms in Real-Time Internets-Of-EveryThings Which Operate in the Big-Squared Data Space," *SoutheastCon 2018*, 2018, pp. 1–9.

[21] K. Hong, D. Lillethun, U. Ramachandran, B. Ottenwalder, and B. Koldehofe, "Mobile Fog: A Programming Model for Large-Scale Applications on the Internet of

Things," in *Proceedings of ACM SIGCOMM Workshop on Mobile Cloud Computing*, 2013, pp. 15–20.

[22] N. K. Giang, M. Blackstock, R. Lea, and V. C. M. Leung, "Developing IoT Applications in the Fog: A Distributed Dataflow Approach," in *Proceedings of the 5th International Conference on the Internet of Things*, 2015, pp. 155–162.

[23] S. Agarwal, S. Yadav, and A. K. Yadav, "An Efficient Architecture and Algorithm for Resource Provisioning in Fog Computing," *International Journal of Information Engineering and Electronic Business*, vol. 8, no. 1, pp. 48–61, 2016.

[24] S. Sarkar, S. Chatterjee, S. Misra, and R. Kudupudi, "Privacy-Aware Blind Cloud Framework for Advanced Healthcare," *IEEE Communications Letters*, vol. 21, no. 11, pp. 2492–2495, 2017.

[25] Y. Cao, P. Hou, D. Brown, J. Wang, and S. Chen, "Distributed Analytics and Edge Intelligence: Pervasive Health Monitoring at the Era of Fog Computing," in *Proceedings of ACM Workshop on Mobile Big Data*, 2015, pp. 43–48.

[26] S. Nandyala and H.-K. Kim, "From Cloud to Fog and IoT-based Real-Time U-Healthcare Monitoring for Smart Homes and Hospitals," *International Journal of Smart Home*, vol. 10, no. 2, pp. 187–196, 2016.

[27] S. Chakraborty, S. Bhowmick, P. Talaga, and D. P. Agrawal, "Fog Networks in Healthcare Application," in *Proceedings of the 13th IEEE International Conference on Mobile Ad Hoc and Sensor Systems*, 2016, pp. 386–387.

[28] J. K. Zao, T. T. Gan, and C. K. You, "Augmented Brain Computer Interaction Based on Fog Computing and Linked Data," in *Proceedings of International Conference on Intelligent Environments*, 2014, pp. 374–377.

[29] A. Monteiro, H. Dubey, L. Mahler, Q. Yang, and K. Mankodiya, "Fit: A Fog Computing Device for Speech Tele-Treatments," in *Proceedings of IEEE International Conference on Smart Computing*, 2016, pp. 1–3.

[30] H. Dubey, J. Yang, N. Constant, A. M. Amiri, Q. Yang, and K. Makodiya, "Fog Data: Enhancing Telehealth Big Data through Fog Computing," *ASE Big Data & Social Informatics*, 2015, pp. 14:1–14:6.

[31] S. K. Datta, C. Bonnet, and J. Haerri, "Fog Computing Architecture to Enable Consumer Centric Internet of Things Services," in *Proceedings of IEEE International Symposium on Consumer Electronics*, 2015, pp. 1–2.

[32] N. B. Truong, G. M. Lee, and Y. Ghamri-Doudane, "Software Defined Networking-Based Vehicular Adhoc Network with Fog Computing," in *Proceedings of IFIP/IEEE International Symposium on Integrated Network Management*, 2015, pp. 1202–1207.

[33] S. Salonikias, I. Mavridis, and D. Gritzalis, "Access Control Issues in Utilizing Fog Computing for Transport Infrastructure," in *Proceedings of International Conference on Critical Information Infrastructures Security*, Springer, 2015, pp. 15–26.

[34] X. Chen and L. Wang, "Exploring Fog Computing Based Adaptive Vehicular Data Scheduling Policies through a Compositional Formal Method – PEPA," *IEEE Communications Letters*, vol. 21, no. 4, pp. 745–748, 2017.

[35] F. Malandrino, C. Chiasserini, and S. Kirkpatrick, "The Price of Fog: A Data-Driven Study on Caching Architectures in Vehicular Networks," in *Proceedings of the 1st*

ACM International Workshop on Internet of Vehicles and Vehicles of Internet, 2016, pp. 37–42.

[36] B. Tang, Z. Chen, G. Hefferman, T. Wei, H. He, and Q. Yang, "A Hierarchical Distributed Fog Computing Architecture for Big Data Analysis in Smart Cities," *ASE Big Data & Social Informatics*, 2015, pp. 28: 1–28:6.

[37] J. Li, J. Jin, D. Yuan, M. Palaniswami, and K. Moessner, "EHOPES: Data-Centered Fog Platform for Smart Living," in *Proceedings of International Conference on Telecommunication Networks and Applications*, 2015, pp. 308–313.

[38] F. Y. Okay and S. Ozdemir, "A Fog Computing Based Smart Grid Model," in *Proceedings of International Symposium on Networks, Computers and Communications*, Yasmine Hammamet, Tunisia, May 2016, pp. 1–6.

[39] M. A. A. Faruque and K. Vatanparvar, "Energy Management-as-a-Service over Fog Computing Platform," *IEEE Internet Things Journal*, vol. 3, no. 2, pp. 161–169, 2016.

[40] I. Abdullahi, S. Arif, and S. Hassan, "Ubiquitous Shift with Information Centric Network Caching Using Fog Computing," *Computational Intelligence in Information Systems*, 2015, pp. 327–335.

[41] W. Lee, K. Nam, H. G. Roh, and S. H. Kim, "A Gateway Based Fog Computing Architecture for Wireless Sensors and Actuator Networks," in *Proceedings of the 18th International Conference on Advanced Communication Technology*, 2016, pp. 1–1.

[42] Y. Xu, V. Mahendran, and S. Radhakrishnan, "Towards SDN-based Fog Computing: MQTT Broker Virtualization for Effective and Reliable Delivery," in *Proceedings of the 8th International Conference on Communication Systems and Networks*, 2016, pp. 1–6.

[43] M. Peng, S. Yan, K. Zhang, and C. Wang, "Fog-Computing-Based Radio Access Networks: Issues and Challenges," *IEEE Network*, vol. 30, no. 4, pp. 46–53, 2016.

[44] M. Aazam and E.-N. Huh, "E-HAMC: Leveraging Fog Computing for Emergency Alert Service," in *Proceedings of IEEE International Conference on Pervasive Computing and Communication Workshops*, 2015, pp. 518–523.

[45] V. K. Sehgal, A. Patrick, A. Soni, and L. Rajput, "Smart Human Security Framework Using Internet of Things, Cloud and Fog Computing," *Intelligent Distributed Computing*, 2015, pp. 251–263.

[46] N. Chen, Y. Chen, Y. You, H. Ling, P. Liang, and R. Zimmermann, "Dynamic Urban Surveillance Video Stream Processing Using Fog Computing," in *Proceedings of the 2nd IEEE International Conference on Multimedia Big Data*, 2016, pp. 105–112.

Chapter 11

Towards a "Green"-er Internet of Things

Due to its applicability in a wide spectrum of domains, cloud computing has gained increasing popularity. The cloud computing framework is typically based on data center networks (DCNs), which are essentially networks of interconnected cloud data centers (CDCs) and are responsible for computation and storage within the cloud. All data processing and management applications are run and service requests served from these CDCs. However, with the introduction of the IoT, there has been a steep increase in the number of Internet-connected devices; thus, the amount of data to be handled by the CDCs is growing every passing day. To put things in perspective, as of 2017, about 1 million statuses, comments, and photos are added to Facebook, about half a million tweets and photos are shared over Twitter and Snapchat, respectively [1]. Concerning service provisioning, Google conducts 3.6 million searches per minute, and The Weather Channel serves more than 18 million forecast requests over the same duration [1]. There are more than 23 billion Internet-connected devices as of 2018. Statista predicts this number to grow over 31 billion by 2020 and up to 75 billion by 2025 [2]. With this increasing trend in the number of edge devices, DCNs face heavy network traffic, which affects service latency significantly. Consequently, applications requesting latency-sensitive services experience a deterioration in the QoS and QoE. Processing this massive amount of data involves a huge migration overhead and presents a big challenge before the cloud in terms of quicker processing of data and serving many applications simultaneously.

The CDCs are required to be up and running around the clock in order to facilitate instantaneous service provisioning for incoming application requests. This consumes a massive amount of energy and emits an enormous amount of greenhouse gases (GHGs), especially carbon dioxide (CO_2) [3], which takes a tremendous toll on the environment. The 30 million servers located across the globe accounted for 100 terawatt-hours of the world's power consumption in 2007, costing more than $9 billion; this statistic is expected to grow up to 200 terawatt-hours in the next few years [4,5]. About 70 billion kilowatt-hours of electricity were consumed in 2014 by the CDCs located within the United States alone [6]. It is, therefore, pivotal to keep the CDCs from being overwhelmed by incoming data streams and service requests. With the integration of the cloud and fog computing models, a major part of the cloud's workload can be moved closer to the network's edge, within the fog computing tier. Processing a big part of the data volume and serving a section of the application requests from the distributed fog tier reduces the workload on the core cloud computing module and improves energy efficiency, overall service latency, and cost.

In this chapter, we analyze the suitability of the fog computing paradigm to serve the demands of real-time, latency-sensitive applications in the context of the IoT. We show the characterization of the two computing paradigms in terms of power consumption and service latency. We discuss the suitability of fog computing in the context of the IoT. With the classical cloud computing service model demanding the CDCs be up and running around the clock, irrespective of the rate of the incoming data streams or service requests, the amount of GHGs emitted in this process becomes unreasonably high. Currently, the total annual electricity consumption for information and communication technology (ICT) purposes account for about 1,500 terawatt hours, which is equal to the total electricity produced by Japan and Germany annually [7]. We show a comparative analysis of the energy efficiency (i.e., the "green"-ness) of the two computing paradigms: fog and cloud. We extend our analysis on the energy efficiency by demonstrating the variation in the power consumption and cost for different renewable and non-renewable energy resources.

11.1 Reference Model

Before we present the latency and energy analysis for the fog computing and classical cloud computing paradigms, we present the reference architecture for our analysis. We first present the assumptions made in the design of our model, and then we discuss in detail the architecture and network model for this analysis.

11.1.1 Assumptions

The fog computing paradigm is still very much in its early days; ideas are primarily theoretical with little practical and commercial-scale development to date. It is, therefore, important to explicitly present the few assumptions made in course of our analysis.

- Fog computing, by definition, provides support for device mobility. The edge devices at the bottom tier are assumed to know their geo-spatial location. These devices are equipped with technologies such as the Global Positioning System, Geographic Information System, and Global Navigation Satellite System and are assumed to periodically report their location information.
- The devices in the fog tier (i.e., the fog devices) are assumed to be either (a) standard networking devices such as routers, gateways, or switches, empowered with additional computing and storage abilities [3,8–10] or (b) miniaturized specialized computing nodes such as the Raspberry Pi and Intel Edison, capable of hosting the data processing environment and supported by a limited amount of storage [11,12].
- The fog tier provides support for device mobility. As edge and IoT devices are mostly mobile, communication between fog instances (FI) or intra-FI communication and information transfer are necessary to provide undisrupted continued service without any unnecessary delays. Inter-FI communication is expected to be present in the system.

11.1.2 Reference Architecture

We now present the fog computing architecture to be used as a reference in this discussion. As shown in Figure 11.1, we consider a hierarchical computing architecture with three tiers: the bottom tier populated by edge or IoT devices, typically mobile in nature; the middle tier, the fog computing tier, where location-specific FIs are present; and the top, where we have the core centralized cloud computing framework.

As discussed in Chapter 10, the edge devices at the bottom tier are clubbed together, based on their spatial location, to form location-based logical groups, known as virtual clusters (VCs). Together, these VCs form an Edge Virtual Private Network (EVPN) [3], through which data streams are transferred from the edge tier to the different FIs in the fog tier. As these edge devices are assumed to be inherently mobile, the mapping of an edge device to an FI is injective and dynamic. Based on the operations performed, the fog computing tier can be classified into two sub-parts: (a) the fog abstraction sub-tier and (b) the fog orchestration sub-tier [9]. The fog abstraction sub-tier is responsible for the management of the fog resources, enabling virtualization within the fog tier, and

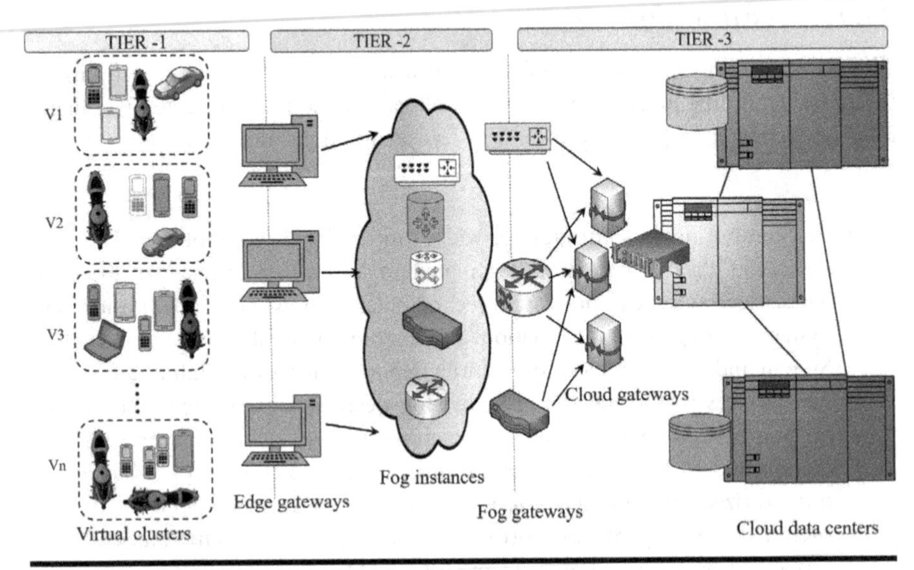

Figure 11.1 Networking links and components of fog computing [3]

preserving data privacy. The fog orchestration sub-tier, on the other hand, holds the exclusive properties of fog computing. The orchestration layer comprises a software agent, known as *foglet,* which monitors the state of the fog devices, manages the distributed databases within the FI to account for scalability and fault tolerance, and hosts a service orchestration module, which is responsible for policy-based routing of incoming data flow. Within the FIs, the orchestration module data pre-processes the service requests and decides whether they need to be transmitted to the CDCs. The service requests that require storage or historical data analysis are re-directed to the CDCs, while the others are processed within the fog tier. The CDCs are equipped with massive computing and storage ability and are responsible for long-term storage and global analysis of the data.

11.2 Networking Model

We now discuss the functional components and the data transfer links associated with the fog computing paradigm as stated in [3]. Communication between the edge and fog tiers takes place over Bluetooth, ZigBee, or WiFi, whereas communication between the fog and cloud tiers often takes place over high-speed wired or wireless connections.

We denote the set of all geo-spatially distributed mobile edge devices by $\mathcal{N}$, where $|\mathcal{N}| = N$ indicates the total number of edge devices present at the edge tier. At any given time, we consider these N edge devices to be grouped into a finite number of VCs, based on their spatial locations. The set of all such VCs is denoted by $\mathcal{V}$, such that $\mathcal{V} = \{v_i | i \in [1, V], i \in \mathbb{I}\}$and $|\mathcal{V}| = V$, where V is the total number of VCs present in the system and v_i denotes the ith VC. The total number of edge devices mapped to v_i is denoted by $n_i, \forall i = 1(1)V$. As we have assumed that at any instant of time, total coverage of the edge devices by the VCs is preserved and that an edge device may belong to only one VC at a time, we have

$$\sum_{i=1}^{V} |v_i| = \sum_{i=1}^{V} n_i = N. \tag{11.1}$$

The data generated by the different edge devices within a VC are transported through the connecting edge gateway towards the fog computing tier. The set of all these edge gateways is denoted by $\mathcal{E}$, with $|\mathcal{E}| = E$ denoting the total number of edge gateways in the system. The data transport links between the VCs and the edge gateways are considered to be unrestricted, varying widely based on the communication technology used by the different edge devices. However, in this analysis, we have considered the data transmission bandwidth between the EVPNs and the FIs to be restricted. The set of all FIs present in the system is given by $\mathcal{F}$ and the total number of FIs denoted by $|\mathcal{F}| = F$. All data and queries generated from the application instances running within the different edge devices are forwarded through the (e,f) link, between the EVPN e and the FI f, $\forall e \in \mathcal{E}$ and $\forall f \in \mathcal{F}$ [3].

Let $P_x^{v_i}(t)$ be the total amount of data (inclusive of the data streams and the service requests), in bytes, which are generated by the edge devices located within the boundary of the VC v_i in a time-slot of duration t. $P_x^{v_i}(t)$ includes the data to be processed, denoted by $P_r^{v_i}(t)$, and the data that are to be simply stored (and processed later in an ad-hoc manner), denoted by $P_s^{v_i}(t)$. Clearly, $P_x^{v_i}(t) = P_r^{v_i}(t) + P_r^{v_i}(t)$. These data include any raw data stream generated by the sensor-equipped edge devices and any service requests generated by the application instances. Based on the type of data stream or the service request, they are to be either processed and served from within the fog tier or forwarded to the cloud computing tier. If an incoming data stream is associated with services that must be served within a short time, the stream and the corresponding service requests are processed and served, respectively, by some FI in the fog tier, without intervention of the cloud tier. However, the requests that require intervention of the cloud computing layer for analysis based on historical data sets and for long-term (semi-permanent or permanent) storage and the corresponding data streams are redirected to the cloud tier periodically, after a time duration of t. From the total data generated by the edge devices within v_i over the duration of t, we assume that $Q_x^{v_i}(t)$ are the total number of data (inclusive of the data streams and the service requests) in bytes that are redirected from the fog tier to the cloud computing layer. Among these, $Q_r^{v_i}(t)$ bytes are to be processed within the cloud, whereas, $Q_s^{v_i}(t)$ bytes of data are to be simply stored. Clearly, $Q_x^{v_i}(t) = Q_r^{v_i}(t) + Q_s^{v_i}(t)$, $P_r^{v_i}(t) \leq Q_r^{v_i}(t)$, *and* $P_s^{v_i}(t) \leq Q_s^{v_i}(t)$, $\forall i = 1(1)V$.

The fog gateways located between the fog and cloud tiers, are represented by the set $\mathcal{G}$, where the total number of fog gateways is represented by $|\mathcal{G}| = G$ [3]. Also, $\forall f \in \mathcal{F}$ and $\forall g \in \mathcal{G}$, the data communication link, (f, g), between FI, f, and the fog gateway, g, is considered to be restricted in terms of network bandwidth. Finally, we present the data communication and data aggregation within the cloud computing tier. The data redirected from the fog tier going through the fog gateways reaches some CDC. This transfer is assumed to take place over a channel with limited bandwidth. The set of all CDCs is represented by $\mathcal{K}$, where $|\mathcal{K}| = K$ denotes the total number of CDCs present in the system. $\forall g \in \mathcal{G}$ and $\forall c \in \mathcal{K}$, the communication link between the cloud gateway, g, and the DC, c, is denoted by (g, c). It is through this set of links that the data reach the cloud tier from the fog tier. However, we assume here that only one CDC is responsible for processing a given data stream at a given instant of time. We consider that at any given time instant, the processing takes place only in one CDC, following the migration of the other required data chunks from the corresponding CDCs. This migration of data is considered to take place over high bandwidth links.

11.3 Performance Metrics

We introduce two new routing variables, one each for the fog and the cloud computing tiers. The fog tier routing variable (FTRV), $X_{v_i, e, f}^{fog}(t)$, essentially keeps track of the route the data streams follow after their origination at some edge device until they reach an FI. Mathematically, $\forall v_i \in \mathcal{V}, \forall e \in \mathcal{E}, \forall f \in \mathcal{F}$, $X_{v_i, e, f}^{fog}(t)$ corresponds to the uplink route through which the data generated from the VC v_i in time-slot t reaches the FI, f [3]. The cloud tier routing variable (CTRV) applies to the data that have been redirected from the fog tier to the cloud tier for purposes such as global-scale aggregation, historical analysis, and long-term storage. The CTRV, denoted by $X_{f, g, c, d}^{cld}(t)$, is defined as the route followed by the data streams from the fog tier to the CDC, where they have been received. Mathematically, $\forall f \in \mathcal{F}, \forall g \in \mathcal{G}, \forall c, d \in \mathcal{D}$, $X_{f, g, c, d}^{cld}(t)$ denotes the route through which data from the FI, f, reaches its destination CDC in time-slot t. Therefore, in a nutshell, the FTRV is the route, $v_i \rightarrow e \rightarrow f$, that represents the path along which the data originating from v_i moves to the FI, f, through the intermediate edge gateway, e, and the CTRV, $f \rightarrow g \rightarrow c \rightarrow d$, indicates that the data stream is redirected from the FI, *f*, to the CDC, *c*, through the intermediate fog gateway, *g*, and from there it is once again migrated to another CDC, *d*, for aggregation and further processing [3].

The magnitudes of the FTRV and CTRV variables corresponding to any given route are computed as the ratio of the volume of data (in bytes) that traverses through the route in duration *t* to the total data (in bytes) generated during that period. Clearly, if a given route for data transmission is valid, and at least some data traverses through the route in a given time duration, the corresponding value of the FTRV or the CTRV is set as non-zero for that duration. Also, as stated in [3], $\sum_{v_i \in \mathcal{V},\, e \in \mathcal{E},\, f \in \mathcal{F}} X^{fog}_{v_i,\, e,\, f}(t) = 1$ implies that every byte of data originating from v_i reaches *f* without any loss, whereas, $\sum_{v_i \in f \in \mathcal{F},\, g \in \mathcal{G},\, c,\, d \in \mathcal{D}} X^{cld}_{f,\, g,\, c,\, d}(t) = 1$ indicates that the amount of data redirected towards CDC, *d*, being re-directed from the FI, *f*, has reached its destination without any loss. At any given time instant, *t*, the set of all feasible FTRVs ($\mathcal{X}^{fog}$) is expressed as

$$\mathcal{X}^{fog} = \{X^{fog}_{v_i,e,f}(t) \mid X^{fog}_{v_i,e,f}(t) = [0,1]; \sum_{v_i \in \mathcal{V}, e \in \mathcal{E}, f \in \mathcal{F}} X^{fog}_{v_i,e,f}(t) = 1,\ \forall v_i \in \mathcal{V}, \forall e \in \mathcal{E}, \forall f \in \mathcal{F}\} \quad (11.2)$$

Similarly, at time, *t*, the set of feasible CTRVs ($\mathcal{X}^{cld}$) is given by

$$\mathcal{X}^{cld} = \{X^{cld}_{f,g,c,d}(t) \mid X^{cld}_{f,g,c,d}(t) = [0,1]; \sum_{v_i \in f \in \mathcal{F}, g \in \mathcal{G}, c, d \in \mathcal{D}} X^{cld}_{f,g,c,d}(t) = 1, \forall f \in \mathcal{F}, \forall g \in \mathcal{G}, \forall c, d \in \mathcal{D}\}. \quad (11.3)$$

Having defined the routing variables, we now describe the role of these two variables in quantification of the different performance metrics related to our analysis.

11.3.1 Power Consumption

The power consumption metric is divided into sub-components for data that are processed within the fog tier without any inference of the cloud computing framework. However, for the data that are processed within the cloud tier, the power consumption metric has one additional sub-component. The additional component included for the cloud tier is due to migration of the data from one CDC to another, whereas the three sub-components common to the fog and cloud tiers are the power consumption due to data forwarding, computation, and storage.

- *Data forwarding*: During the phenomenon of data forwarding from one tier to another, power is consumed due to several operations such as transmission of the data, reception of the data, and initial processing required for routing of the data packets. Together, these attributes contribute to the *power consumption due to data forwarding*, denoted by $\Psi_{df}^{fog}(t)$, for power consumption over a duration of t. The notation $\sum_{v_i,e,f}$, as used in the chapter, essentially means the same as the notation $\sum_{v_i\in\mathcal{V},e\in\mathcal{E},f\in\mathcal{F}}$. For data forwarding from the edge tier to the fog tier, $\Psi_{df}^{fog}(t)$ is computed as

$$\begin{aligned}\Psi_{df}^{fog}(t) &= (\gamma_{eg}+\gamma_{fi})[\sum_{i=1}^{V}\{P_r^{v_i}(t)-Q_r^{v_i}(t)+P_s^{v_i}(t)-Q_s^{v_i}(t)\}\sum_{v_i,e,f}X_{v_i,e,f}^{fog}(t)] \\ &= (\gamma_{eg}+\gamma_{fi})[\sum_{v_i,e,f}\{P_r^{v_i}(t)-Q_r^{v_i}(t)+P_s^{v_i}(t)-Q_s^{v_i}(t)\}X_{v_i,e,f}^{fog}(t)],\end{aligned} \tag{11.4}$$

where γ_{eg} and γ_{fi} represent the amount of energy required per second (power) to forward one unit byte of data by the edge gateways and the FIs, respectively [3].

Similarly, for the data streams forwarded from the fog tier to the cloud tier over a duration of t, the corresponding power consumption due to data forwarding is denoted by $\Psi_{df}^{cld}(t)$. Once again, the notation $\sum_{f,g,c,d}$ is used as an alternative to the notation $\sum_{v_i\in f\in\mathcal{F},g\in\mathcal{G},c,d\in\mathcal{D}}$ in the chapter. For data forwarding from the fog tier to the cloud tier, $\Psi_{df}^{cld}(t)$ is expressed as

$$\begin{aligned}\Psi_{df}^{cld}(t) &= (\gamma_{eg}+\gamma_{fi})[\sum_{i=1}^{V}\{Q_r^{v_i}(t)+Q_s^{v_i}(t)\}\sum_{v_i,e,f}X_{v_i,e,f}^{fog}(t)] \\ &\quad+\gamma_{cl}[\sum_{i=1}^{V}\{Q_r^{v_i}(t)+Q_s^{v_i}(t)\}\sum_{f,g,c,d}X_{f,g,c,d}^{cld}(t)] \\ &= \sum_{i=1}^{V}\{Q_r^{v_i}(t)+Q_s^{v_i}(t)\}[(\gamma_{eg}+\gamma_{fi})\sum_{v_i,e,f}X_{v_i,e,f}^{fog}(t) \\ &\quad+\gamma_{cl}\sum_{f,g,c,d}X_{f,g,c,d}^{cld}(t)],\end{aligned} \tag{11.5}$$

where γ_{cl} is the power required to forward one unit byte of data by a cloud gateway.

- *Computation*: Based on the type of services with which the data streams are associated, the processing locations of the streams are determined.

Computations, therefore, take place both at the fog tier and at the cloud. Data computation within the fog tier is primarily to support the real-time service requests and temporal data storage. The FIs are therefore equipped with a limited amount of storage, and the data is periodically moved to the cloud storage. Let τ denote the time-to-live for every data packet after which it is removed from the fog storage and moved to the cloud. Therefore, over an interval of time, t, the power consumption due to computation within the fog layer ($\Psi_{cp}^{fog}(t)$) depends on the data stored within the FIs from time $(t-\tau)$ till the present (we assume τ to be shorter than t) [3]. Mathematically, ($\Psi_{cp}^{fog}(t)$) is expressed as

$$\Psi_{cp}^{fog}(t) = \beta^{fog} \sum_{j=t-\tau}^{t} \phi_j^{fog} \sum_{i=1}^{V} \{P_s^{v_i}(j) - Q_s^{v_i}(j)\}, \tag{11.6}$$

where β^{fog} is the mean power consumption per byte due to computations and is computed as the ratio of the power exhausted for the processing of an instruction to the number of bytes in the instruction. ϕ_j^{fog} is the weight factor associated with the data set to be processed, and the magnitude of ϕ_j^{fog} decreases with the increase in the age of the data. The magnitude of ϕ_j^{fog} lies within (0, 1), with $\phi_j^{fog} = 1$ for $j = t$. The term $\sum_{i=1}^{V} \{P_s^{v_i}(j) - Q_s^{v_i}(j)\}$ represents the cumulative amount of data stored within the temporary fog storage.

The computations and subsequent analysis on the data are rather extensive in the CDCs, both in terms of the volume of data to be processed and the complexity of the analysis. Over a duration of t, the total power consumption due to computation within the CDC, denoted by $\Psi_{cp}^{cld}(t)$, is a cumulative function of the amount of data stored within the CDCs, beginning from $t = 0$. Mathematically, ($\Psi_{cp}^{cld}(t)$) is computed as

$$\Psi_{cp}^{cld}(t) = \beta^{cld} \sum_{j=0}^{t} \phi_j^{cld} \sum_{i=1}^{V} Q_s^{v_i}(j), \tag{11.7}$$

where β^{cld} is the mean power required to process one unit byte at the CDC, and $\sum_{i=1}^{V} Q_s^{v_i}(j)$ is the total amount of data aggregated within a CDC for computation, with $\phi_{ij}^{cld} \in [0, 1]$.

- *Storage*: Similar to the computational power consumption, the power consumption due to storage depends on the volume of data (number of bytes) stored. The mathematical expression for the storage power consumption for t duration for the fog tier ($\Psi_{st}^{fog}(t)$) is given by

$$\Psi_{st}^{fog}(t) = \alpha^{fog} \sum_{j=0}^{t} \sum_{i=1}^{V} \{P_s^{v_i}(j) - Q_s^{v_i}(j)\}. \tag{11.8}$$

Similarly, we can compute the power consumption due to data storage in the CDCs ($\Psi_{st}^{cld}(t)$), and it is expressed as

$$\Psi_{st}^{cld}(t) = \alpha^{cld} \sum_{j=0}^{t} \sum_{i=1}^{V} Q_s^{v_i}(j), \tag{11.9}$$

where α^{fog} and α^{cld} represent the power consumed due to data storage per byte per unit time, within the databases in the fog storage unit and the CDC, respectively [3].

While the three aforementioned power factors are applicable for both the fog and cloud tiers, the next factor is applicable only to the cloud.

- *Data migration*: Data analysis at the cloud computing tier invariably requires migration of the different data sets distributed across the different CDCs. Therefore, the power consumption due to data migration refers to the total power consumed to migrate the required data from the different CDCs to the aggregator CDC. The aggregator CDC may change with the types of data involved in an analysis and type of analysis to be performed on the data. At time t, the overall migration cost ($\Psi_{mg}^{cld}(t)$) within the cloud computing framework is given by [3]

$$\Psi_{mg}^{cld}(t) = \begin{cases} \sum_{c \in \mathcal{D}} \sum_{d \in \mathcal{D}} \eta_{cd} \sum_{j=0}^{t-1} \phi_j^{cld} \sum_{i=1}^{V} Q_s^{v_i}(j), & \text{if } \mathcal{A}_t \neq \mathcal{A}_{t-1} \\ 0 & , \text{otherwise} \end{cases} \tag{11.10}$$

where η_{cd} is power consumption per byte for migration of data from CDC $c, d \in \mathcal{D}$. $\mathcal{A}_t$ is essentially the aggregator CDC chosen for the integration over the time duration t. Therefore, $\mathcal{A}_t = \mathcal{A}_{t-1}$ implies that the aggregator CDC remains unchanged over two consecutive intervals, which indicates that no additional migration is required for the second integration.

Total power consumption: Therefore, over a time duration of t, the overall power consumption within the fog tier, $\Psi^{fog}(t)$, is expressed as

$$\Psi^{fog}(t) = \Psi^{fog}_{df}(t) + \Psi^{fog}_{cp}(t) + \Psi^{fog}_{st}(t). \quad (11.11)$$

With an additional factor, the power consumption in the cloud tier, $\Psi^{cld}(t)$, is given by

$$\Psi^{cld}(t) = \Psi^{cld}_{df}(t) + \Psi^{cld}_{cp}(t) + \Psi^{cld}_{st}(t) + \Psi^{cld}_{mg}(t). \quad (11.12)$$

11.3.2 Service Latency

Like power consumption, the latency incurred in service provisioning can be listed under two sub-heads: latency incurred due to data transmission and latency caused by data processing. Service latency is the turnaround time for an application request to be successfully served. Unrestricted communication bandwidths between the edge devices and the EPVN and between the different CDCs make the corresponding service latencies negligible. Thus, the bottleneck in the network is formed at the juncture of the fog and cloud tiers, formed by the links (e, f) and (f, g), $\forall e \in \mathcal{E}, \forall f \in \mathcal{F}, \forall g \in \mathcal{G}$ [3]. As the network bandwidth between the edge gateways and the FIs, $\mathbb{W}(e,f)$, is restricted, with the increase in the number of active edge devices and simultaneous service requests, service latency increases. The communication links between the fog gateways and the cloud gateways are bandwidth constrained ($\mathbb{W}(f,g)$) as well. The two constituent sub-heads for the service latency are separately discussed.

- *Transmission latency*: Let us assume δ_{ef} and δ_{fg} to be the latencies incurred in transmitting a unit byte of data from an EVPN to an FI and from a fog gateway to a cloud gateway, respectively. Therefore, the latency for the data to traverse from the edge tier to the fog computing tier, computed over a time duration of t ($\delta^{fog}_{tr}(t)$), is expressed as

$$\delta^{fog}_{tr}(t) = \delta_{ef} \sum_{i=1}^{V} \{P^{v_i}_{r}(t) + P^{v_i}_{s}(t) - Q^{v_i}_{r}(t) - Q^{v_i}_{s}(t)\}. \quad (11.13)$$

Similarly, for requests that are redirected by the fog tier towards the cloud computing tier, the corresponding transmission latency ($\delta^{cld}_{tr}(t)$) is expressed as

$$\delta_{tr}^{cld}(t) = (\delta_{ef} + \delta_{fg}) \sum_{i=1}^{V} \{Q_r^{v_i}(t) + Q_s^{v_i}(t)\}. \tag{11.14}$$

Therefore, with the presence of the fog computing tier between the edge and the cloud core, the mean transmission latency over t, (Δ_{tr}^{fog}), is computed as

$$\Delta_{tr}^{fog}(t) = \frac{\delta_{ef} \sum_{i=1}^{V} \{P_r^{v_i}(t) + P_s^{v_i}(t)\} + \delta_{fg} \sum_{i=1}^{V} \{Q_r^{v_i}(t) + Q_s^{v_i}(t)\}}{\sum_{i=1}^{V} \{P_r^{v_i}(t) + P_s^{v_i}(t)\}} \tag{11.15}$$

In contrast, in case of the traditional cloud computing framework, where the fog tier is absent, the mean transmission latency, ($\Delta_{tr}^{cld}(t)$), is given as

$$\Delta_{tr}^{cld}(t) = \frac{\delta_{eg} \sum_{i=1}^{V} \{P_r^{v_i}(t) + P_s^{v_i}(t)\}}{\sum_{i=1}^{V} \{P_r^{v_i}(t) + P_s^{v_i}(t)\}}, \tag{11.16}$$

where δ_{eg} is the latency for transmission of a single byte of data from an edge device to the CDC. Clearly, from the triangle inequality, we have $\delta_{eg} \geq \delta_{ef} + \delta_{fg}$.

- *Processing latency*: The latency incurred due to processing of data prior to serving a request is categorized as processing latency. Formally, the processing latency is defined as the time required to serve a request after processing and analyzing the corresponding data sets that have been accumulated over the previous τ duration. The processing latency within the fog computing tier, at time t ($\delta_{pr}^{fog}(t)$), is expressed as

$$\delta_{pr}^{fog}(t) = (P_r^{v_i}(t) - Q_r^{v_i}(t))\zeta^{fog} \sum_{j=t-\tau}^{t} \phi_j^{fog} \sum_{i=1}^{V} \{P_s^{v_i}(j) - Q_s^{v_i}(j)\}. \tag{11.17}$$

Similarly, for the services requests that are redirected to the cloud computing layer, the processing latency is defined in terms of the time required to aggregate the migrated data sets from the different CDCs, followed by the processing time within the aggregator CDC. The processing latency ($\delta_{pr}^{cld}(t)$) within the cloud tier is given as

$$\delta_{pr}^{cld}(t) = Q_r^{v_i}(t)\zeta^{cld} \sum_{j=t-\tau}^{t} \phi_j^{cld} \sum_{i=1}^{V} Q_s^{v_i}(j), \tag{11.18}$$

where ζ^{fog} and ζ^{cld} represent the per-byte processing latency at the fog computing and cloud computing tiers, respectively. The mean processing delay at time t for a fog computing environment is, therefore, expressed as follows:

$$\begin{aligned} \Delta_{pr}^{cld}(t) =& [(P_r^{v_i}(t) - Q_r^{v_i}(t))\zeta^{fog} \sum_{j=t-\tau}^{t} \phi_j^{fog} \sum_{i=1}^{V} \{P_s^{v_i}(j) - Q_s^{v_i}(j)\} \\ &+ Q_r^{v_i}(t)\zeta^{cld} \sum_{j=t-\tau}^{t} \phi_j^{cld} \sum_{i=1}^{V} Q_s^{v_i}(j)] / \sum_{i=1}^{V} \{P_r^{v_i}(t) + P_s^{v_i}(t)\}. \end{aligned} \tag{11.19}$$

On the contrary, the mean transmission latency, $\Delta_{pr}^{cld}(t)$, at time t for a classical cloud computing environment is given as

$$\Delta_{pr}^{cld}(t) = \frac{P_r^{v_i}(t)\zeta^{cld} \sum_{j=t-\tau}^{t} \phi_j^{cld} \sum_{i=1}^{V} P_s^{v_i}(j)}{\sum_{i=1}^{V} \{P_r^{v_i}(t) + P_s^{v_i}(t)\}}. \tag{11.20}$$

Having shown the expressions for both the transmission latency and service latency, in order to compute the mean service latency, we can simply add the mean transmission latency and the mean processing latency for the respective computing framework [3].

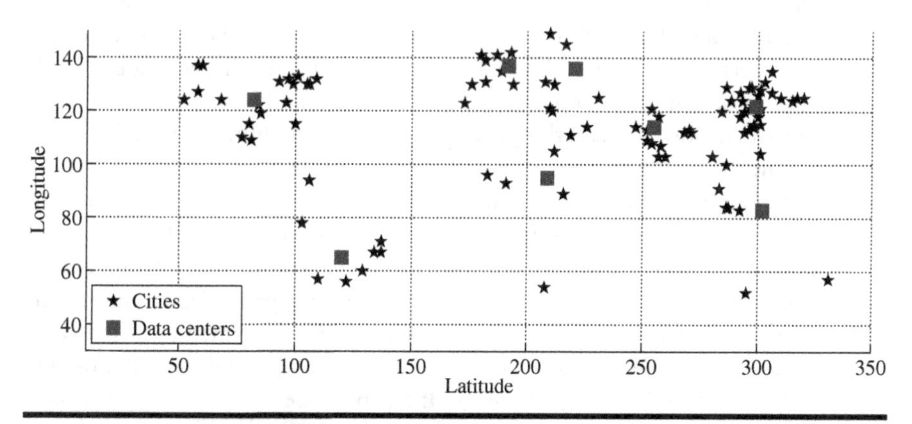

Figure 11.2 Global deployment scenario [3]

11.4 A Case Study: Simulation Setup

In the following two sections, we present our analysis of the fog computing paradigm in terms of power consumption, CO_2 emission rate, and cost. The CO_2 emission rate acts as the measure of the "green"-ness of the computing paradigm, in the context of the IoT. We also provide a comparative analysis of fog computing and the classical cloud computing paradigm with respect to these metrics. Before we show these analyses in the following section, we explain the simulation setup used to produce the results.

11.4.1 Network Topology

The three fundamental components of the fog computing paradigm are the set of edge devices ($\mathcal{N}$) at the bottom tier, the set of FIs ($\mathcal{F}$) in the middle tier, and the set of CDCs ($\mathcal{D}$) at the top of the hierarchy.

- *Edge devices*: We consider a global-scale deployment of these mobile nodes. We take the list of the 100 most populated cities around the globe [13], along with the proportion of the population that use Internet-related services [14], and the geographic location (longitude and latitude) [15] of the respective cities. Based on these data, we plot the points in the simulation, as shown in Figure 11.2, and constitute a 100×100 matrix, $L_c[1..100][1..100]$, which stores the relative Euclidean distance between any two cities [3]. The edge devices within a particular city are logically grouped to form a VC. Due to device mobility and random switch-on/switch-off patterns of the edge devices, we consider that the total number of edge devices may vary within the range [10,000, 100,000]. This will also allow us to access the system performance against variable network conditions. The number of edge devices in each city is taken proportionally based on the number of the Internet users in the city.
- *Fog instances*: For each of the 100 cities, we assume the existence of one city-specific FI. In a real scenario, based on the traffic generated by the edge devices in a city, there may be more than one FI per city. However, in this discussion, to keep things simple, we keep the number for all cities set to one.
- *Cloud data centers*: In our example, we consider 8 CDCs are spread across the global landscape, the locations of which are determined based on the population distribution among the 100 cities considered for our simulation. The 8×8 matrix, $L_d[1..8][1..8]$, stores the pair-wise Euclidean distances between the different CDCs. Every CDC is assumed to host a varied number of IT components within the discrete set {16,000, 32,000, 64,000, 128,000}, based on the network traffic it processes [3].

11.4.2 Network Traffic

We assume that the data generated from the cities is proportional to the number of Internet-service users for the respective cities. The data streams are transmitted in the form of packets from the VCs through the EVPNs to the fog tier. The sizes of the data packets vary from 34 bytes (size of packet header and FCS) to 65,550 bytes (maximum payload size). Each instruction is considered 64 bits. The packets arrive in a pattern following the Poisson distribution, with a mean packet arrival rate of 1 packet per edge device per second. The capacity of the (e,f) links, $\forall e \in \mathcal{E}, \forall f \in \mathcal{F}$, is taken as 1 Gbps, and that for the $(f,g), \forall f \in \mathcal{F}, \forall g \in \mathcal{G}$ links, is taken as 10 Gbps [3].

11.4.3 Performance Metrics

We now show the modeling of the different performance metrics used in our analysis [3].

- *Power consumption*: The overall power consumption is computed as the summation of the power consumed by different network, computing, and storage elements in the system. The power consumed by each 1 Gbps and 10 Gbps router is taken as 20 W and 40 W, respectively [16]. Each of the 3-layer network switches and storage components is assumed to consume 350 W and 600 W of power, respectively. Also, within an FI, the total power consumed by the fog computing nodes is taken to be 3.7 W, whereas the power consumed by the CDCs is considered to be proportional to the number of IT elements present in each CDC, which falls within the range {9.7, 19.4, 38.7, 77.4} MW.
- *CO_2 Emission*: The amount of CO_2 gas emission is directly dependent on the source of the energy. In Table 11.1, we present the CO_2 emission rate for the different energy sources, which are used for our analysis [17]:
- *Cost*: In this analysis, cost is assessed in terms of both operational cost and the CO_2 emission penalty for the different component elements. The cost of each 1 Gbps and 10 Gbps router port is taken as $50/year, whereas the cost of a server is $4,000/year [16]. The cost to upload 1 byte of data is considered to be $12, and to store 1 GB of data, the cost is assumed to uniformly vary within the range $0.45 to $0.55/hour. The electricity cost is taken as uniformly distributed between $30/MWh and $70/MWh [16]. However, the CO_2 emission penalty is considered to have a higher impact on the environment, and the penalty is taken to be $1,000/ton CO_2 emitted.

Table 11.1 CO_2 emission rates [3]

Type of energy source	*Energy source*	*CO_2 emission rate (in g/kWh)*
Non-renewable	Coal	960
	Diesel	778
	Natural gas	443
Renewable	Geothermal	38
	Hydroelectric	10
	Wind	9

11.5 A Case Study: Performance Evaluation

The objective of this analysis is to assess the suitability and applicability of fog computing as a potential platform to support the IoT. We investigate the eco-friendliness of the fog computing paradigm in this context and compare it to that of the classical cloud computing paradigm. While we have shown the modeling of the various aspects of the fog computing framework for this analysis, the cloud computing-related parameters are excerpted from the work of Zhang *et al.* [18].

11.5.1 Service Latency

As mentioned earlier, the service latency is computed as a sum of the transmission latency and the processing latency [3]. We first discuss the analysis of the variation of the service latency with the number of edge devices at the bottom-most tier. We define the ratio of the total bytes transmitted to the fog computing tier to the number of bytes referred to the cloud computing core as the *cloud transmission ratio*, mathematically represented by Θ. We show the plot of the transmission latency and propagation latency for both fog computing and the conventional cloud computing architectures with the change in the magnitude of Θ. The value of Θ is varied within the range [0.05, 0.75], and the corresponding change in the service latencies is observed.

As shown in Figures 11.3(a) and (b), for different magnitudes of Θ, we show the plots of the mean transmission latency and mean processing latency against the variable number of edge devices. We observe that as the magnitude of Θ decreases (i.e., as more services demand real-time and latency-sensitive services), the mean transmission latency and the mean processing latency diminishes accordingly. Also, we note that with the increase in the number of active edge devices, the transmission and processing latencies increase.

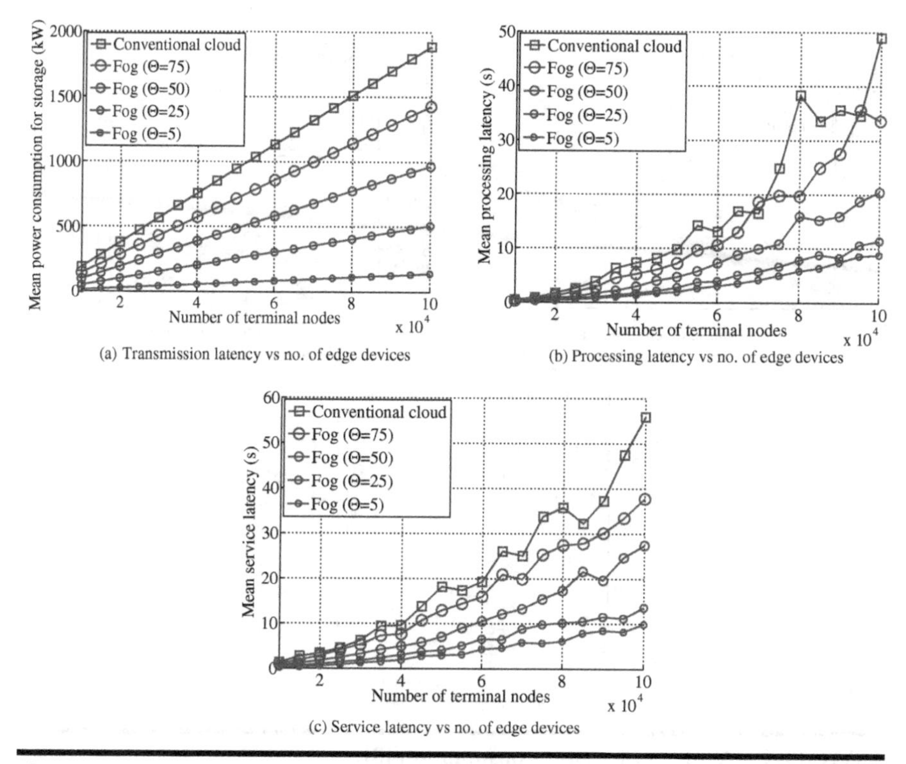

Figure 11.3 Analysis of service latency [3]

The variation of the overall service latency for both these computing paradigms, however, are observed to follow the same pattern. To summarize, if the percentage of the data or service requests that demand real-time services is low (i.e., if Θ has a higher magnitude (Θ ≃ 100)), the service latency for fog computing becomes almost the same as that for a cloud computing environment. Fog computing, in such cases, acts as an additional intermediate tier that serves no purpose, as all the data streams and service requests are eventually redirected to the cloud tier.

11.5.2 Power Consumption

We now show the analysis of the power consumption, considering the individual contributions of the different constituent factors such as data forwarding, computation, and storage [3]. In Figure 11.4(a), we observe that the mean power consumption increases linearly with the increase in the number of active edge devices in the network, the impact of the variation in the magnitude of Θ is

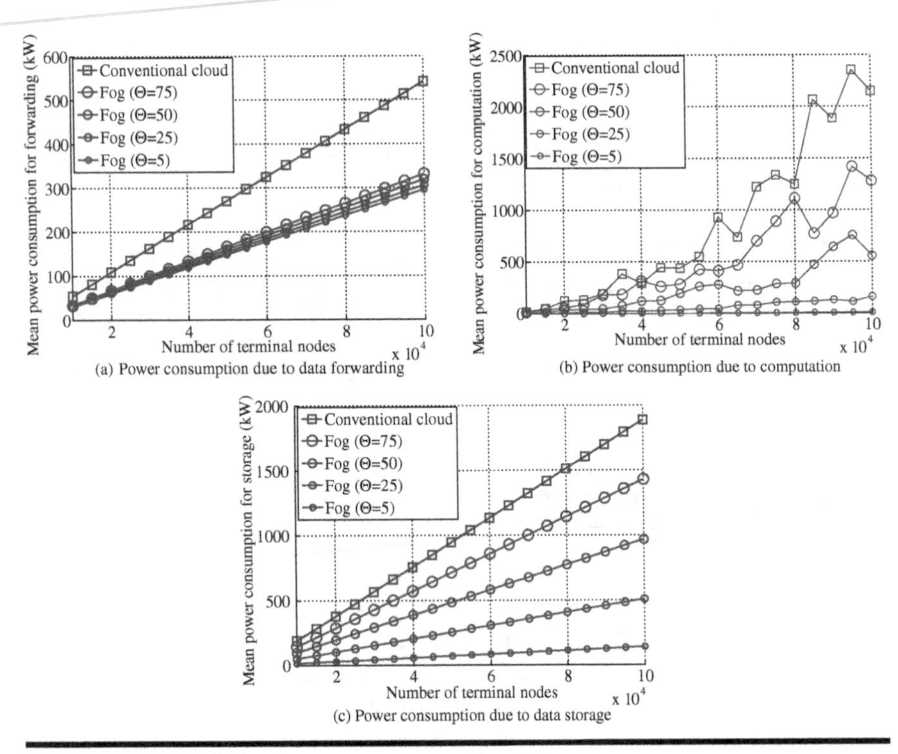

Figure 11.4 Analysis of power consumption [3]

noted to be very low. However, in comparison with classical cloud computing, the mean power consumption is significantly less. The variation in the mean power consumption due to computation and storage with the change in the number of active edge devices is shown in Figures 11.4(b) and 11.4(c), respectively. Similar inferences are drawn in these cases, as the power consumption in both cases is noticed to decrease significantly as the magnitude of Θ decreases. Also, in the presence of the fog computing tier, the overall mean power consumption, for $\Theta = 75$, is computed to be 42.2%, as compared to the classical cloud-based computing framework.

11.5.3 CO_2 Emission

We now move to the heart of our analysis: the effects of the two computing paradigms on the environment in terms of the amount of CO_2 gas emitted [3]. We divide this analysis based on the source of the energy, as we consider both renewable and non-renewable sources.

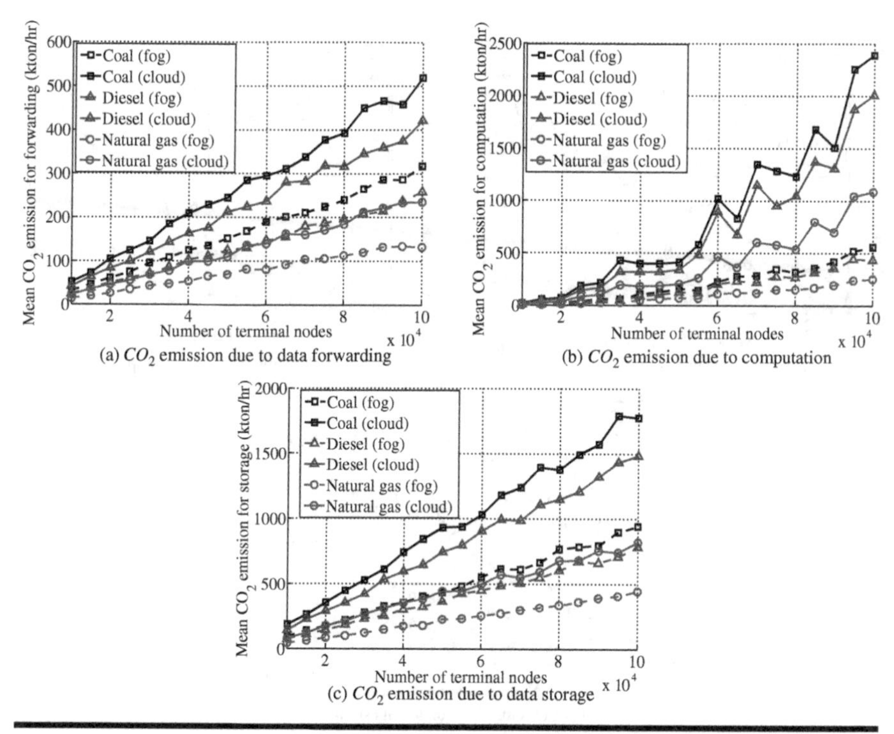

Figure 11.5 Analysis of CO_2 emission for non-renewable energy sources [3]

In Figure 11.5, we plot the variation in the mean CO_2 emission against the number of active edge devices for each of the contributing factors (i.e., data forwarding, computation, and data storage). As shown in Figure 11.5(a), the mean CO_2 emission for forwarding data packets is notably high in cloud computing paradigm compared to fog computing when non-renewable energy sources such as coal, diesel, and natural gas are considered. For the same energy sources, in Figures 11.5(a) and 11.5(c), we observe that fog computing, once again, emerges as the greener computing framework when separate computation and storage factors are taken into account, respectively.

For the renewable energy resources, we show a similar analysis, as shown in Figure 11.6. The three renewable sources considered in this analysis are geothermal, hydroelectric, and offshore wind. Similar to Figure 11.5, it is observed that fog platforms are distinctly more energy efficient in terms of CO_2 emissions, considering packet forwarding (Figure 11.6(a)), computation (Figure 11.6(b)), and storage (Figure 11.6(c)) [3].

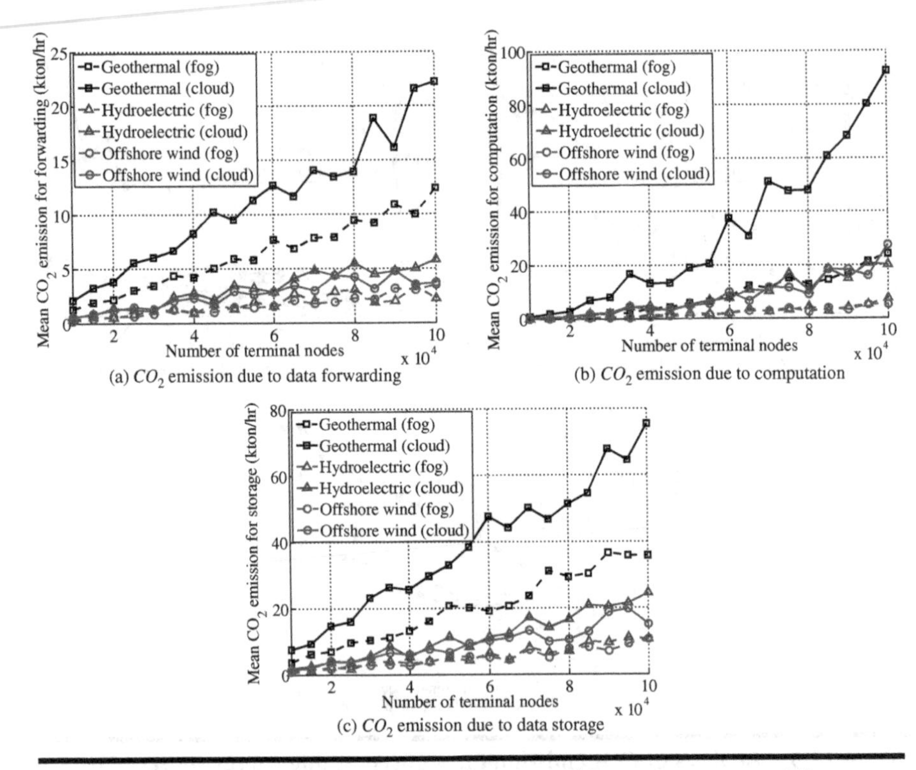

Figure 11.6 Analysis of CO_2 emission for renewable energy sources [3]

11.5.4 Cost

In the final part of our analysis, we discuss the cost for fog and cloud. We take into account the penalty for CO_2 gas emission for the computing nodes both at the fog and cloud tiers and show the computation of the overall cost accordingly. It is observed that for both non-renewable, as shown in Figure 11.7(a), and renewable energy sources, as shown in Figure 11.7(b), the fog-based systems exhibit a cheaper nature. Also, with the use of renewable energy, even in the worst-case scenario, the cost is reduced by at least half with the introduction of the renewable energy sources.

11.6 Summary

In this chapter, we have discussed how the fog computing paradigm plays a significant role in "green"-ing the IoT. We first presented the reference model and the architecture. Next, we presented the networking model. We showed how

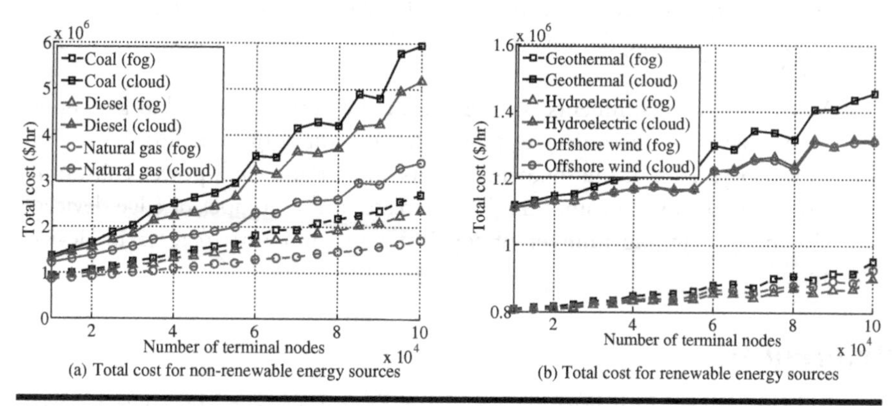

Figure 11.7 Analysis of cost for non-renewable and renewable energy sources [3]

the performance metrics (power consumption and service latency) can be modeled. We demonstrated the scenario with a case study in which we discussed in detail the network topology, the network traffic, and the performance metrics. Followed by this, we further demonstrated a case study on performance evaluation.

Working Exercises

Numerical Problems

1. If the total amount of data to be processed in the cloud is 2 GB, and the fog computation weight factor is 0.8, then what will be the total power exhausted due to computation over the said amount of data, given that the power required to process a unit byte of data is 10 mW?
2. What is the value of the cloud computation weight factor, if the power exhausted to process 100 MB data at the cloud data center is 2×10^5 W and the mean power consumption to process a unit byte of data is 5 mW?
3. If the power consumed to store a unit byte of data in the fog storage is 3 mW, then how much power is required to store 1 GB of data?
4. If the power consumption due to data forwarding, computation, and storage in the fog tier is 2 kW, 3 kW, and 1 kW, respectively, then what is the total power consumed in the fog tier?

Conceptual Questions

1. Define and differentiate between CTRV and FTRV.
2. What are the contributing factors for power consumption for the cloud and fog tiers?

3. What are the two contributing factors for computation of service latency?
4. Which component do you think will play a bigger part and under what condition?
5. What are your comments on the use of the various renewable and non-renewable energy resources to power the fog computing platform?
6. How do you think the following events affect the overall power consumption and cost in fog computing? (i) increase in the number of edge devices, (ii) increase in the number of fog instances, and (iii) increase in device mobility.

References

[1] B. Marr, "How Much Data Do We Create Every Day? The Mind-Blowing Stats Everyone Should Read," *Forbes*, Online: https://tinyurl.com/ycydna2e, 2018.

[2] "8 Statistics that Prove IoT Will Become Massive from 2018," *Medium*, Online: https://tinyurl.com/y8t9pk3d, 2018.

[3] S. Sarkar, S. Chatterjee, and S. Misra, "Assessment of the Suitability of Fog Computing in the Context of Internet of Things," *IEEE Transactions on Cloud Computing*, vol. 6, no. 1, pp. 46–59, 2018.

[4] D. Bouley, "Estimating a Data Centers Electrical Carbon Footprint," *Schneider Electric – Data Center Science Center*, 2010.

[5] "Report to Congress on Server and Data Center Energy Efficiency Public Law 109 - 431," *U.S. Environmental Protection Agency ENERGY STAR Program*, Online: https://tinyurl.com/yb37zjb6, 2007.

[6] Y. Sverdlik, "Here's How Much Energy All US Data Centers Consume," *DataCenter Knowledge*, Online: https://tinyurl.com/y96cy9rb, 2016.

[7] M. P. Mills, "The Cloud Begins With Coal," *Digital Power Group*, 2013.

[8] S. Sarkar and S. Misra, "Theoretical Modeling of Fog Computing: A Green Computing Paradigm to Support IoT Applications," *IET Networks*, vol. 5, no. 2, pp. 23–29, 2016.

[9] F. Bonomi, R. Milito, P. Natarajan, and J. Zhu, "Fog Computing: A Platform for Internet of Things and Analytics," N. Bessis and C. Dobre (eds.), *Big Data and Internet of Things: A Roadmap for Smart Environments, Studies in Computational Intelligence*, Springer, vol. 546, 2014. Online: https://link.springer.com/chapter/10.1007/978-3-319-05029-4_7

[10] K. Hong, D. Lillethun, U. Ramachandran, B. Ottenwalder, and B. Koldehofe, "Mobile Fog: A Programming Model for Large-Scale Applications on the Internet of Things," in *Proceedings of the 2nd ACM SIGCOMM Workshop on Mobile Cloud Computing*, New York, USA, 2013, pp. 15–20.

[11] D. Borthakur, H. Dubey, N. Constant, L. Mahler, and K. Mankodiya, "Smart Fog: Fog Computing Framework for Unsupervised Clustering Analytics in Wearable Internet of Things," in *Proceedings of the IEEE Global Conference on Signal and Information Processing (Global SIP)*, Montreal, Canada, 2017, pp. 472–476.

[12] Y. Elkhatib, B. Porter, H. B. Ribeiro, M. F. Zhani, J. Qadir, and E. Riviere, "On Using Micro Clouds to Deliver the Fog," *IEEE Internet Computing*, vol. 21, no. 2, pp. 8–15, 2017.

[13] T. Brinkoff, "City Population," Online: www.citypopulation.de/, 2013.

[14] "Internet World Stats," Online: www.internetworldstats.com/, 2013.

[15] "Google Map," Online: www.google.com/maps.

[16] F. Larumbe and B. Sanso, "A Tabu Search Algorithm for the Location of Data Centers and Software Components in Green Cloud Computing Networks," *IEEE Transactions on Cloud Computing*, vol. 1, no. 1, pp. 22–35, 2013.

[17] B. K. Sovacool, "Valuing the Greenhouse Gas Emissions from Nuclear Power: A Critical Survey," *Energy Policy*, vol. 8, no. 36, pp. 2950–2963, 2008.

[18] L. Zhang, C. Wu, Z. Li, C. Guo, M. Chen, and F. C. M. Lau, "Moving Big Data to the Cloud: An Online Cost-Minimizing Approach," *IEEE Journal on Selected Areas in Communications*, vol. 31, no. 12, pp. 2710–2721, 2013.

Chapter 12

Security in the IoT

12.1 Introduction

In the previous chapters, we covered the vision of the IoT, as well as enabling technologies, including how they can be realized in real life, technically and algorithmically. Currently, the IoT is in its nascent stage; like any other market, it will grow if the developers, innovators, and leaders are able to resolve market challenges. In a survey conducted collaboratively with McKinsey and the Global Semiconductor Alliance (GSA), it was discovered that security in the IoT is a major concerns of its beneficiaries [1]. The importance of security has been repeatedly pointed out by hackers who attempted to infiltrate IoT devices. Examples of such incidents include Mirai Botnet (aka Dyn Attack) in 2016, one of the largest denial-of-service attacks [2], the Stuxnet attack, which sabotaged the uranium enrichment facility in Iran [3], and attacks by Wikileaks, which stole more than 8,500 documents from the Central Intelligence Agency by exploiting iOS and Android vulnerabilities [4].

12.1.1 IoT Security vs. Conventional Security

Before we delve into the details of security in the IoT, it is important to understand the contextual significance and to figure out how IoT security is very different from conventional security in wireless networks or the cloud. The primary difference begins with the deployment of the network. Unlike other wireless networks, IoT networks are deployed on low power and lossy networks (LLNs), which are characterized by their limited memory and processing power, thereby mandating the use of lightweight protocols and discarding heavy in-network optimizations. Node impersonation is another major challenge in LLNs (i.e., if an attacker is able to connect using a fake identity, it is considered authentic, which should be prevented) [5]. In IoT devices, security issues are explored at different layers. Frequency hopping-based communication and public key cryptography are impossible at the perception layer, owing to the constrained behavior of the nodes. The network layer is vulnerable to man-in-the-middle and counterfeit attacks. Provision for data sharing creates a wide spectrum of security concerns due to privacy policies, access control, and lack of confidentiality of information. For example, Datagram Transport Layer Security is used in the network layer, IPv6 is used in the perception layer, Constrained Application Protocol (CoAP) is used in the application layer, and so on.

12.1.2 Difference between Security and Privacy

There is usually some confusion regarding the difference between *security* and *privacy*, as there are no formalized or standard definitions. A commonly asked question is "*what is the key difference between data privacy and data security*?" We will explore the answer to this question before we move into subsequent topics. These terms are related; in some cases, they behave synonymously. Let us take the example of our home [6]. We would like our home to be *secure*. To ensure this, we generally keep the doors locked or give the keys only to trustworthy people. This helps to control access to our home. If we would like to have *privacy* at our place (which most people do), we would not allow people to enter our bedrooms and bathrooms. A particular environment can be private but not necessarily secure, for example showrooms in supermarkets and malls. On the other hand, areas of airports close to the boarding gates are secure but not private.

From the perspective of data, any publicly readable data are considered secure if they cannot be modified without due permission; however, this does not make the data necessarily private. For example, if personal data is sold for marketing purposes, we may not be comfortable with it, although it was not stolen. On the other hand, private data handled insecurely would include leakage of bank details, credit card information, and so on. We would state that while security focuses on controlled and authorized access to data to protect the infrastructure and network from cyber attacks, data privacy is the right of the owner to restrict the outward flow of data [7].

12.2 Security in the IoT

The IoT comprises three different layers: application, perception, and network, as shown in Figure 12.1 [5].

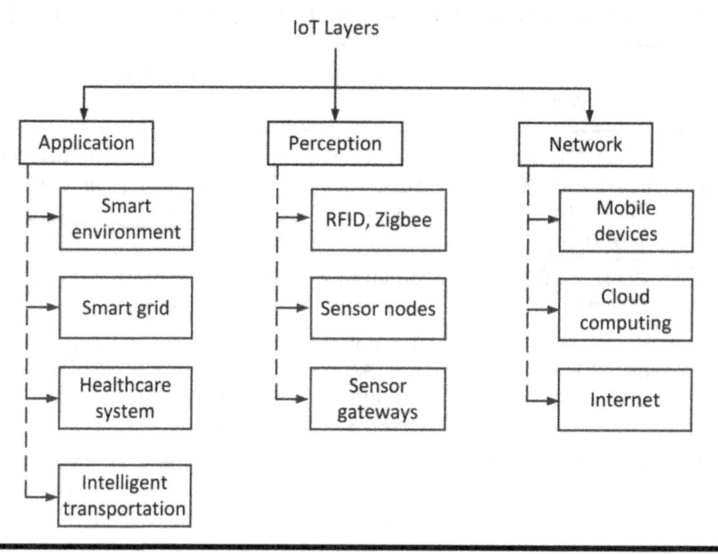

Figure 12.1 Layers of IoT

- ***Application layer***: This is the uppermost layer, which is visible and accessible to end-users. It contains applications such as smart environment monitoring, smart grid, and healthcare applications. They follow a distributed application layer protocol for data exchange and communication. The protocol uses a middleware to communicate with the service provider platform (such as cloud) through M2M or D2D. Table 12.1 illustrates the vulnerabilities of this layer.
- ***Perception layer***: This layer is mainly responsible for collection of information from data sources. This layer is divided into two parts – a perception node, which includes the data sources, and the perception network, which is further connected to the IoT network layer [5]. It supports a number of communication protocols including RFID, ZigBee, and Bluetooth.
- ***Network layer***: The network layer is responsible for providing network security during transmission of data through layers. It includes (i) mobile devices such as phones, laptops, and smart vehicles, which are potentially vulnerable to attacks (e.g., eavesdropping, corruption, and denial of service), (ii) distributed Internet-based computing infrastructures such as the cloud, and (iii) the Internet.

Table 12.1 Potential security threats at application layer of the IoT [5]

Applications	*Modes of communication*	*IoT devices*	*Types of threat*
Smart city applications	Wi-Fi, Ultra-wide-band, ZigBee, Bluetooth, LTE, 6LoWPAN	Environment sensors, home appliances	Authentication, privacy, eavesdropping, denial-of-service
Smart grid	Wi-Fi, ZigBee, Z-Wave	Smart meters and smart readers	Privacy, eavesdropping, tampering
Smart healthcare solutions	Bluetooth, ZigBee, IEEE 802.15.6 protocol	Wearable body area sensors, ambulatory sensors	Privacy, authentication authorization, denial-of-service, distributed denial-of-service
Smart vehicular system	WiFi, Bluetooth	Smart cars, road-side units, parking sensors	Jamming, congestion, security, spectrum sharing

12.2.1 Protocol Stack for the IoT

The IoT follows a protocol stack, which is still evolving and under research. The main communication protocols are as follows [8]:

- ***IEEE 802.15.4*** is a low-power communication protocol, mainly used for communication at the physical layer. It also lays the foundation for communication at the higher layers of the IoT protocol stack.
- ***6LoWPAN*** is used for packet fragmentation and subsequent transmission of the fragmented packets using IPv6. It enables direct communication with nodes operating in other low-energy protocols at the higher layers.
- ***Routing Protocol for Low Power and Lossy Networks (RPL)*** facilitates the routing of 6LoWPAN to the IoT application through optimized communication architecture.
- ***Constrained Application Protocol (CoAP)*** facilitates inter-operability within heterogeneous devices with varied communication standards and data formats.

The security requirements of the aforementioned protocols are shown in Figure 12.2. The IEEE 802.15.4 protocol deals with the time synchronization and access control within the network. In 6LoWPAN, there is no dedicated security management. The RPL standard has a separate field and a higher order bit of that field to denote security specifications. The DLTS protocol is related to the confidentiality and integrity of the data in the application layer along with CoAP.

12.2.2 Security Threats in the IoT

IoT modules are vulnerable to various threats. We list the different types of attacks in Table 12.2 [9].

- Network infrastructure: Attacks are possible within different IoT elements, from wireless networks to the mobile nodes and cloud/fog DCs.
  - Denial of Service (DoS) attacks are among the most popular in communication networks. These attacks have restricted effects on the infrastructure. Core infrastructure suffering from distributed DOS attacks is partially affected, as the IoT comprises autonomous and semi-autonomous protocols, which carry out the necessary functions.
  - Man in the Middle includes eavesdropping and malicious traffic injection. For example, if gateways within the networks are attacked or compromised, it is possible for the attacker to gain access to the interfaces, thereby modifying or corrupting the data or virtual machines.
  - Rogue Gateway: Many times, end-user devices enable deployment of gateways by attackers or malicious users. This is similar to the Man in the Middle attack, as it opens opportunities for eavesdropping and injecting traffic.
- Edge DCs: Edge DCs are key elements for IoT service infrastructures; they act as the intermediate hubs of computation for latency-sensitive applications. Edge DCs have publicly accessible APIs for end-users, cloud providers, VMs, and web applications. DCs are highly vulnerable owing to the exposure of their interfaces.
  - Physical damage constitutes external damage to the geographically distributed fog/edge nodes, which are relatively small in size. This requires an attacker to be physically close to the device. However, it is very likely that such attacks are witnessed by common people around that region.

Table 12.2 Security threats in IoT elements [9]

Challenge	*Description*
Infrastructure	Interoperability; monitoring; accountability
Virtualization	VM lifecycle; container and context awareness
Resources & tasks	Resource location; task scheduling; offloading
Distribution	Cooperation; n-tier management; soft state
Mobility	Connectivity; seamless handoff
Programmability	Usability; session management

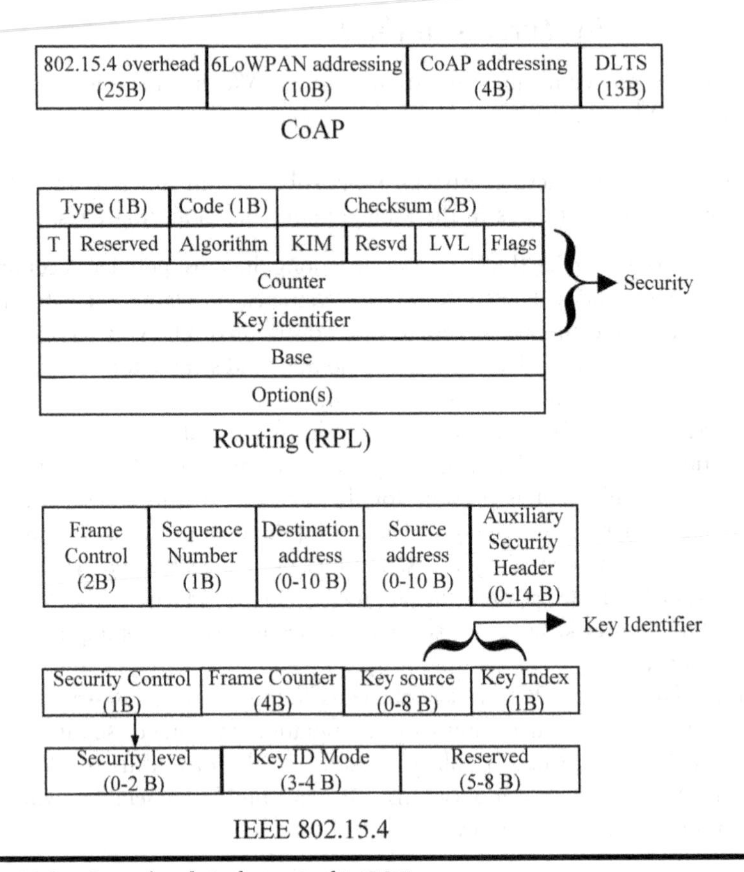

Figure 12.2 Security data format of IoT [8]

- Privacy leakage involves information discharge from the edge DCs, which leads to exposure of local information affecting applications of that region. However, information can sometimes migrate across distributed DCs for resource management or device-mobility management. In such cases, the information leakage may affect other applications remote from the region of attack.
- Privilege escalation occurs due to lack of technical training and experience on the part of the persons managing the edge devices and the fog nodes. This leads to misconfiguration or low security maintenance within devices and exposes them to external attacks. However, the attacks can also be led by internal adversaries.
- Service manipulation happens when an attacker gains access to the internal node and can modify or corrupt privileges, thereby attempting

access to more data for bigger attacks. This may lead to DoS attacks or data services and information tampering.
  - Rogue DC means that through privileged escalation or privacy leakage, an adversary can gain complete control of the infrastructure with full access to data. The adversary may also have the ability to manipulate and control information flow across VMs of other DCs.
- Core DCs are the computational hub of centralized cloud services. For some cases, these core infrastructures are governed by a single organization that also controls the edge infrastructure. However, in some cases, the edge infrastructure service can be delegated to another organization.
  - Privacy leakage happens because information from cloud DCs flows to edge DCs and may be accessed by internal or external adversaries. However, information from the upper layers is likely to be a subset of information at the lower edge layers.
  - Service manipulation means that information flow from cloud to fog nodes may be compromised not only by corrupting the information but also by adding fake or irrelevant information generated from historic data. However, there is a less likelihood of affecting the entire ecosystem with such attacks.
  - Rogue infrastructure occurs when some of the specialized core services can be attacked. Attackers will try to gain access to all layers of cloud and fog, thereby attempting to collapse the system and the network.
- Virtualization infrastructure is responsible for the creation and management of VMs serving end-users. However, such infrastructures can be attacked in various ways. Also, there are possibilities of corrupting or compromising individual VMs on an end-user basis.
  - Denial of Service (DoS): A misbehaving or compromised VM may delete, modify, or corrupt the records of end-users. It may also aggressively request cloud resources with a goal towards depletion. The worst is an attack on edge nodes, as they are inherently resource-constrained.
  - Misuse of resources: Malicious VMs may execute heavy, unwanted, and resource-hungry jobs just to keep the resources occupied. They might also run programs for hacking passwords and making illegal file transfers.
  - Privacy leakage: As mentioned earlier, edge infrastructures contain a large number of publicly accessible APIs, which, if not protected, will have their sensitive information exposed. This may affect the surrounding DCs.
  - Privilege escalation: A malicious VM may try to corrupt and introduce vulnerabilities inside its host. This has a huge spectrum of outcomes ranging from DC collapse, resource blockage and isolation, migration of confidential information, and lost data integrity.

- VM manipulation: Manipulation of VMs may corrupt the access privileges of file system and affect job schedules as well. An adversary can inject malicious codes inclusive of logic bombs, iterative virus attacks, and false encryption of data files.

- User devices: Devices such as phones, laptops, and wearable objects are not only recipients of several IoT services; they also actively participate in contributing data. Malicious users may want to disrupt the services, however, with limited or temporary effects.
  - Injection of information: A user device may intentionally start transmitting fake information at a high rate. It might report erroneous or incompatible values that may affect applications at different levels.
  - Service manipulation: If a device provides services, it can alter the beneficiaries of the service and may corrupt the outcome as well.

12.3 Misbehavior in M2M Communication

Machine-to-machine (M2M) communication is considered an enabler of the IoT. M2M is an autonomous manner of machine manageable communication among several things. The primary goal is to share information between machines with little or no human intervention. M2M communication involves both wired and wireless modes of communication with the control center. Most M2M communications contain machines, which collect information regarding an event and report to the application via the network. At the data center (DC), the information is translated into meaningful action, which, in turn, is triggered through an actuator [10]. M2M communication is composed of several networking machines and gateways. This communication can be described as a three-stage process: data collection, data transmission, and data processing [11]. Data collection is capturing data regarding physical events, transmission of data is sending data to the external server, and processing data refers to the analysis and enforcement of appropriate actions.

Generic M2M architecture is shown in Figure 12.3 [12]. Several nodes, which act as data collectors, are grouped into a small subnetwork, which reports the information to the gateways. These gateways are connected to the backbone network of the external server, where the data are analyzed and further decisions are taken. The increase in the large-scale deployment of several M2M devices leads to the generation of large volumes of data within the DCs, which requires enormous computation power to process. On the other hand, cloud servers provide large computational and storage services at reduced price and manpower. Combining M2M with cloud improves efficiency of data processing and communication at reduced cost and maintenance. Cloud servers provide computational

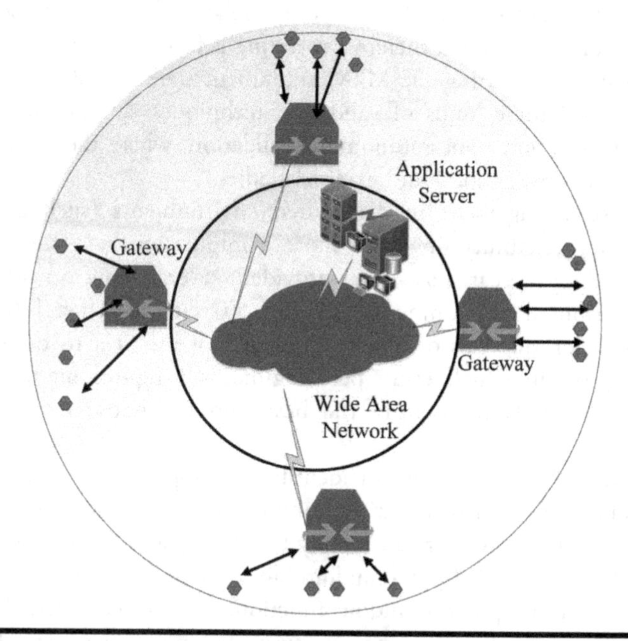

Figure 12.3 Architecture of M2M Communication [12]

services (SaaS) in a distributed fashion and at reduced cost of administration and maintenance.

On the contrary, the mobile nature of M2M devices pose several security threats [13] including physical attacks on devices, compromised credentials, configuration attacks, protocol attacks, attacks on core networks, and user data integrity and privacy attacks. In this section, we will discuss how attacked machines can be safely identified, allowing the rest of the operation to continue without interruption, with the help of dynamic trust management policies.

12.3.1 Where Do We Stand?

With the deployment of M2M technology and its integration with the cloud environment, new research scopes addressing challenges and difficulties have come up. In the existing literature, Osawa [14] explained the problems regarding the global integration of machines to a cloud to provide virtual services, as well as network integration problems regarding global connectivity of communication providers. He also addressed the vertical integration of a network and its terminals to cloud. Cha *et al.* [15] focused on the trust at the equipment level provided by trusted environment in maintaining relevant hardware level security, which forms the root of trust by providing isolation between the software and sensitive data of the machine, thereby protecting it from unauthorized access. The important issue is validating the M2M equipment state for safe operations. The authors described three kinds of validation techniques: autonomous validation, remote validation, and semi-autonomous validation, where the entire validation procedures run between local and external bodies.

Using trust management in distributed environments (such as cloud) for providing data reliability involves more computation, as trust is context-dependent. Many researchers have provided several ways to identify trust levels in distributed environment, including (i) cuboids trust [16], in which trust computation depends on the contribution of the peer to the system, and (ii) Antrep [16], in which every peer maintains a reputation table and uses probabilistic methods to identify the best hop to choose in a transmission path.

Du [17] described the process of identifying compromising sensor nodes in a heterogeneous sensor network with a compromised node detection model. The work assumes that an attacker can modify the data space of the sensor, which can be safeguarded with the help of a random block memory traversal scheme and the noise generated using pseudo random function. Li *et al.* [18] described the data transmission quality as a function for suspected node behavior. If a node is suspected once, a voting scheme is initiated to collect information from the peer nodes. However, most of the works do not detect misbehavior at a particular time instant, which is crucial.

12.3.2 Problem Scenario

We discuss a system that includes a wide variety of machines/devices/nodes ranging from sensor nodes to mobile vehicular agents. The devices/nodes are assumed to report captured event data to the control center with the help of gateways or base stations. At the control centers, a deviation value is calculated from the captured data for every node. Following this, a particular node is suspected of misbehaving. Once the suspected nodes are identified, voting theory is applied, in which information is collected from the non-suspicious nodes regarding the suspected node by votes. Once the voting process is completed, a social aggregation function is used. The output of the aggregation function declares a node as faulty if it is chosen as a compromised node by a majority of voters.

Figure 12.4 illustrates a diagram of the scenario. The machines are represented by nodes, which communicate with the control centers or DCs through the gateways.

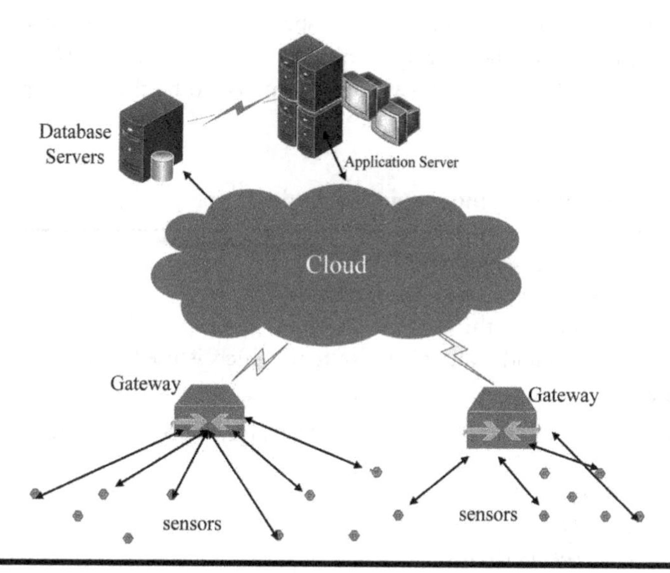

Figure 12.4 Diagrammatic representation of the system scenario

12.3.3 System Model

The algorithm for detection of compromising nodes in M2M communication is executed at the control center. As the algorithm is resource intensive, it cannot be executed at the node level. The assumptions of the solution follow.

- All sensor nodes are static (i.e., the coordinates of the position of the sensors is known *a priori*).
- The sensor nodes are not equipped with tamper-proof hardware.
- Communication between the nodes and/or gateway is assumed bidirectional.

The implementation is expected to run in two stages: (i) suspecting the compromised node and (ii) confirmation of the suspected nodes using the *Theory of Social Choice.*

12.3.4 Operation

The process of suspecting the compromised node involves calculating each node's deviation value from the expected one, which is generally a median calculated over the data of the "n" nodes in the network. Let the number of nodes/devices that are deployed for a particular purpose (e.g., surface temperature reading, fire alarm systems in the forest, and vehicular speed tracking) be "n". Each node is uniquely identified, and the data transmitted by these nodes are D_1, D_2..., D_n. The mode calculated over the data values is assumed to be the expected value for

Algorithm 1 Identifying the suspicious nodes

Inputs:

- D_i: Data values corresponding to each node
- D_b: Mean value over the data values of the nodes
- *Threshold*: Threshold value of the system, which is used to declare a node as a suspect

Output: *suspect*[1..*n*]: A one-dimensional boolean array is used to denote the suspicious nodes.

1: **for** i = 1 to n **do**
2:     $diff = D_b - D_i$
3:     **if** $diff > Threshold$ **then**
4:         $suspect[i] = true$
5:     **else**
6:         $suspect[i] = false$
7:     **end if**
8: **end for**

---

calculating the deviation values. Suspicious nodes are identified with the help of these deviation levels using Algorithm 1.

Once the suspected nodes are identified, the voting process is initiated from the remaining nodes of the network. The voting process works as follows: Let k be the number of suspicious nodes. The voting decision is taken by the remaining $t = n - k$ nodes for the suspecting nodes. We define $Mean[t]$ as a one-dimensional real array containing the mode/mean over the values of the node "i" over t instants of time. Let $Rank[k][t]$ be a two-dimensional array containing the mapping values, which indicates the deviation level of the suspecting nodes from the mean values of the remaining non-suspicious nodes. The range of the rank matrix is a mapping, which defines the level of deviation from its median value, which can be described as:

$$Rank[i][j] = \begin{cases} 1, & \text{if } deviation \leq \Delta \\ 2, & \text{if } \Delta \leq deviation \leq 2\Delta \\ 3, & \text{if } 2\Delta \leq deviation \leq 3\Delta \end{cases} \tag{12.1}$$

The values of Δ depend on the maximum allowable deviation from the system perspective. Once the order of choices is mapped in the *Rank* matrix, the social aggregation function (SAF) [19] is fed with the votes from the remaining nodes. The output of the SAF indicates the aggregated decision about the misbehaving node. Therefore, the SAF is expressed as follows:

$$F(i) = SAF(Rank[i][1], Rank[i][2], \ldots, Rank[i][n]) \qquad (12.2)$$

The work of the SAF is described as a step-by-step process. After obtaining the ranking levels, the weighted mean value of the rank is calculated for each suspicious node, where the weighing factor is the remaining residual battery power of the node. If the obtained mean value is the same as the mode of the rank, then the aggregated decision declares it as the compromising node. If there is a tie, it is resolved by moving the node to a safe state (i.e., ignoring its presence for some threshold period). Thus, we have,

$$F(i) = \begin{cases} 1, & \text{if aggregated decision is faulty} \\ 0, & \text{otherwise} \end{cases} \qquad (12.3)$$

12.3.5 Quantitative Results

The aforementioned algorithm is implemented and executed in a sensor network, which monitors temperature within a certain area with a limited number of sensors (n-nodes). The network has been configured in a way that k out of n nodes start misbehaving ($k << n$). The nodes are uniquely identified and temperature readings are assumed to be recorded in the centigrade scale. The experimental data, shown in Table 12.3, is considered for the performance analysis.

Let us assume that we would like to identify the compromised node at a particular instant, t_{100}. A graph is plotted in Figure 12.5 against the node identifier on the X-axis and the corresponding temperature readings in centigrade on the Y-axis at t_{100}. The nodes in blue are normal, whereas those in red are misbehaving.

In Figure 12.5, we observe that nodes 2, 7, and 10 are identified as suspicious nodes. The mean values of the temperature over the given times are shown in Table 12.3. The graph plotted against nodes to see relative measures is shown in Figure 12.6.

In the confirmation mode, suspicious node behavior is analyzed against its past behavior and is compared with the rest of the trusted nodes for the final detection.

The graph is plotted against the mean temperature (in centigrade) observations of the all nodes on the Y-axis against that of the suspicious nodes on the X-axis.

The threshold range is considered to be [-5, 5]. Therefore, nodes beyond [21, 31] are declared misbehaving or compromising. Based on this logic, nodes 2 and 7 exhibit misbehavior; node 10 deviates negligibly from the mean value of the trusted nodes. Thus, nodes 2 and 7 are confirmed as compromising nodes. Having calculated the mean values, the Rank matrix is fed to the SAF. Considering $\Delta = 2.5$, we have:

$$Rank[i][j] = \begin{cases} 1, & \text{if } deviation \leq 2.5 \\ 2, & \text{if } 2.5 < deviation \leq 5 \\ 3, & \text{if } 5 \leq deviation \leq 7.5 \end{cases} \tag{12.4}$$

Therefore, we have:

$$\begin{pmatrix} 3 & 3 & 3 & 3 & 3 & 3 & 3 \\ 3 & 3 & 3 & 3 & 3 & 3 & 3 \\ 2 & 2 & 2 & 2 & 2 & 2 & 2 \end{pmatrix}$$

At this point, we note the residual power of the nodes in Table 12.5. The table verifies and supports the consideration of the compromising nodes.

Detection of misbehaving nodes can help the network administrators to eliminate the nodes from the network, and thereby, reduce its vulnerability of

Table 12.3 Temperature values (in centigrade) of nodes over various times

Node ID	*Time*							
	t_{10}	t_{20}	t_{30}	t_{40}	t_{50}	t_{70}	t_{85}	t_{100}
N1	27°	29°	24°	21°	25°	27°	24°	29°
N2	47°	42°	14°	22°	9°	15°	67°	52°
N3	25°	26°	27°	24°	26°	27°	26°	26°
N4	24°	27°	26°	25°	25°	26°	26°	29°
N5	27°	24°	27°	23°	25°	27°	26°	25°
N6	26°	25°	22°	24°	25°	26°	24°	25°
N7	9°	3°	5°	77°	57°	47°	66°	52°
N8	28°	24°	23°	25°	26°	27°	25°	26 °
N9	26°	25°	27°	26°	25°	24°	26°	25°
N10	25°	26°	24°	25°	26°	26°	27°	57°

Table 12.4 Mean of temperature readings (in centigrade) at t_{10} and t_{100}

Node ID	*Temperature value*
N1	25.25°
N2	33.5°
N3	25.875°
N4	26°
N5	25.5°
N6	24.625°
N7	39.5°
N8	25.5°
N9	25.5°
N10	29.5°

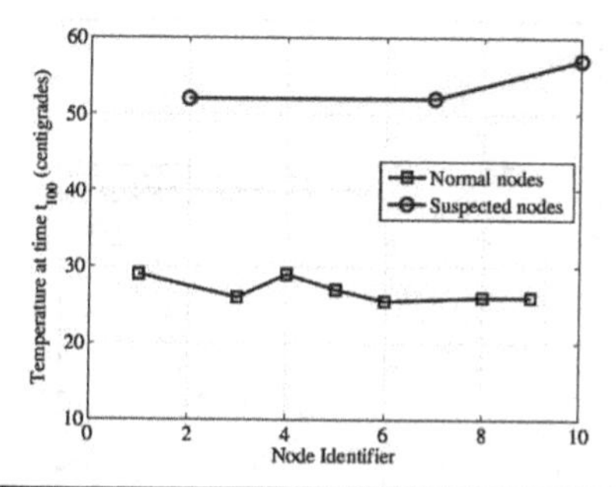

Figure 12.5 Diagrammatic representation of normal and suspicious nodes

the network to external threats. The problems concerning data privacy protection, are however, difficult to quantify in mathematical terms, and often, an intervention from the legislative front is needed [20].

12.4 Summary

In this chapter, we introduced the readers to the concepts of IoT security. We began by stating the differences between IoT security and conventional security challenges in WSNs or the cloud. We also differentiated between the terms "security" and "privacy" in this context. We then presented the protocol stack and security threats in the IoT, followed by a discussion on misbehavior in M2M communication and its impact on the IoT.

Table 12.5 Residual battery power

Node ID	*Residual Battery(%)*
N1	27
N2	45
N3	33
N4	67
N5	34
N6	56
N7	12
N8	33
N9	26
N10	95

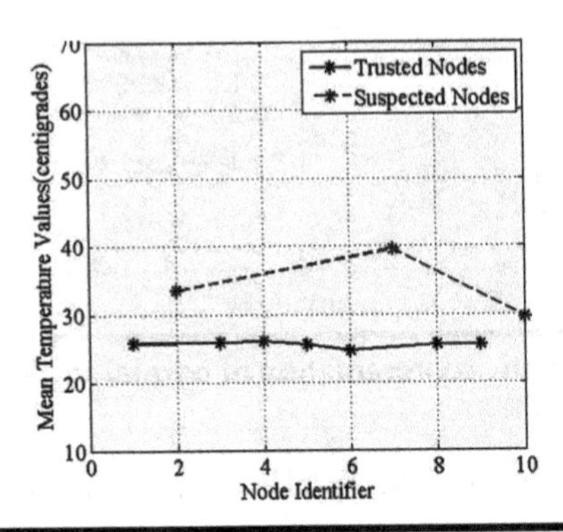

Figure 12.6 Mean of Temperature readings at t_{10} and t_{100}

Working Exercises

Multiple Choice Questions

1. Which of the following is not an IoT layer?
   a) Application
   b) Perception
   c) Analysis
   d) Network

2. Which of the following are the possible security threats for a healthcare IoT application – (i) DoS, (ii) eavesdropping, (iii) authentication authorization, or (iv) tampering?
   a) (i) and (iii)
   b) (iii) and (iv)
   c) (ii) and (iii)
   d) (i), (ii), and (iv)
3. The Constrained Application Protocol (CoAP) focuses on
   a) physical layer communication
   b) packet fragmentation and subsequent transmission in the network layer
   c) inter-operability within heterogeneous devices
   d) data packet routing
4. Which of the following are valid communication mediums in the perception layer of the IoT?
   a) RFID and Zigbee
   b) RFID and Bluetooth
   c) Zigbee and Bluetooth
   d) All of the above
5. Privacy leakage within IoT systems essentially means
   a) corrupting the records of end-users
   b) hacking passwords and making illegal file transfers
   c) intentionally transmitting fake information at a very high rate
   d) exposing sensitive information through publicly accessible APIs

Conceptual Questions

1. How are security threats in the IoT focused around the edge DCs?
2. How does the protocol stack of the IoT facilitate multiple communication mediums?
3. What are the major differences between security and privacy?
4. What are the major differences between IoT security and conventional security?
5. Explain the packet format and the significance of the different fields in IEEE 802.15.4.

References

[1] H. Bauer, O. Burkacky, and C. Knochenhauer, "IoT Security: A Role for Semiconductor Companies," 2017. Online: www.mckinsey.com/industries/semiconductors/ourinsights/security-in-the-internet-of-things.

[2] N. Woolf, "DDoS Attack that Disrupted Internet Was Largest of Its Kind in History, Experts Say," 2016. Online: www.theguardian.com/technology/2016/oct/26/ddos-attack-dyn-miraibotnet.

[3] D. Gewirtz, "Special Report: Stuxnet May Be the Hiroshima of Our Time," 2011. Online: www.zdnet.com/article/special-report-stuxnet-may-be-the-hiroshima-ofour-time/.

[4] L. H. Newman, "WikiLeaks Just Dumped a Mega-Trove of CIA Hacking Secrets," 2017. Online: www.wired.com/2017/03/wikileaks-cia-hacks-dump/.

[5] F. A. Alaba, M. Othman, I. A. T. Hashem, and F. Alotaibi, "Internet of Things Security: A Survey," *Journal of Network and Computer Applications*, vol. 88, pp. 10–28, Jun. 2017, doi: 10.1016/j.jnca.2017.04.002.

[6] E. Zuckerberg, "What Is the Difference between Privacy and Security?" 2017. Online: www.quora.com/What-is-the-difference-between-privacy-and-security.

[7] Techopedia, "What Is the Difference between Security and Privacy?" Online: www.techopedia.com/7/29722/security/what-is-the-difference-between-security-and-privacy.

[8] A. Tiwari and B. B. Gupta, "Security, Privacy and Trust of Different Layers in Internet-Of-Things (Iots) Framework," *Future Generation Computer Systems*, May 2018, doi: 10.1016/j.future.2018.04.027.

[9] R. Roman, J. Lopez, and M. Mambo, "Mobile Edge Computing, Fog Et Al.: A Survey and Analysis of Security Threats and Challenges," *Future Generation Computer Systems*, 2016, doi: 10.1016/j.future.2016.11.009.

[10] I. Stojmenovic, "Machine-To-Machine Communications with In-Network Data Aggregation, Processing, and Actuation for Large-Scale Cyber-Physical Systems," *IEEE Internet of Things Journal*, vol. 1, no. 2, pp.122–128, Apr. 2014.

[11] Y. Zhang, R. Yu, S. Xie, W. Yao, Y. Xiao, and M. Guizani, "Home M2M Networks: Architectures, Standards, and Qos Improvement," *IEEE Communications Magazine*, vol. 49, no. 4, pp. 44–52, Apr. 2011.

[12] M. Booysen, J. Gilmore, S. Zeadally, and G. Rooyen, "Machine-To-Machine (M2M) Communications in Vehicular Networks," *KSII Transactions on Internet and Information Systems*, vol. 6, no. 2, pp. 529–546, 2012.

[13] L. Vesa, "Feasibility Study on Remote Management of USIM Application on M2M Equipment," 3*rd* Generation Partnership Project, Technical Report, May 2007.

[14] T. Osawa, "Practices of M2M Connecting Real World Things with Cloud Computing," *Fujitsu Science & Technology*, vol. 47, no. 4, pp. 401–407, 2011.

[15] I. Cha, Y. Shah, A. Schmidt, A. Leicher, and M. Meyerstein, "Trust in M2M Communication," *IEEE Vehicular Technology Magazine*, vol. 4, no. 3, pp. 69–75, Sept. 2009.

[16] M. Firdhous, O. Ghazali, and S. Hassan, "Trust Management in Cloud Computing: A Critical Review," *CoRR*, abs/1211.3979, 2012.

[17] X. Du, "Detection of Compromised Sensor Nodes in Heterogeneous Sensor Networks," in *Proceedings of IEEE International Conference on Communications (ICC)*, May 2008, pp. 1446–1450.

[18] T. Li, M. Song, and M. Alam, "Compromised Sensor Nodes Detection: A Quantitative Approach," in *Proceedings of the 28th International Conference on Distributed Computing Systems Workshops (ICDCS)*, Jun. 2008, pp. 352–357.
[19] D. Bouyssou, T. Marchan, and P. Perny, Decision-Making Process: Concepts and Methods, Hoboken, NY, USA: Wiley, ISBN: 978-1-848-21116-2, Jul. 2009.
[20] S. Sarkar, J.-P. Banâtre, L. Rilling, and C. Morin, "Towards Enforcement of the EU GDPR: Enabling Data Erasure," in *Proceedings of IEEE International Conference of Internet of Things*, pp. 1–8, 2018.

Index

D

E

F

G

H

I

L

M

N

P

Q

R

S

T

U

V

W